Practical Experimental Designs and Optimization Methods for Chemists

Distribution: VCH Verlagsgesellschaft mbH, P.O. Box 1260/1280, D-6940 Weinheim,
   Federal Republic of Germany

USA and Canada: VCH Publishers, Inc., 303 N.W. 12th Avenue, Deerfield Beach, FL 33442–1705, USA

Practical
Experimental Designs and
Optimization Methods
for Chemists

Charles K. Bayne and Ira B. Rubin

Charles K. Bayne
Oak Ridge National Laboratory
Oak Ridge, Tennessee

Ira B. Rubin
Oak Ridge National Laboratory
Oak Ridge, Tennessee

Library of Congress Cataloging-in-Publication Data

Bayne, C. K. (Charles Kenneth), 1944–
  Practical experimental designs and optimization
methods for chemists.

  Includes bibliographies and index.
  1. Chemistry—Experiments. 2. Experimental design.
I. Rubin, Ira B., 1921–   .   II. Title.
QD43.B38   1986   507′.8   86-9046
ISBN 0-89573-136-3

© 1986 VCH Publishers, Inc.

Printed in the United States of America.

ISBN 0-89573-136-3   VCH Publishers
ISBN 3-527-26195-8   VCH Verlagsgesellschaft

PREFACE

The purpose of this book is to introduce statistically designed experiments to chemists who conduct experiments for the purpose of making inferences from data. To achieve this goal, we emphasize statistical considerations for preliminary planning of experiments, standard statistical designs that may be used for experiments, and the underlying logic for using these designs. It is a common but major error to view statistics as a tool to be used only after experiments are completed. Even using their most sophisticated tools, statisticians who receive data from improperly designed experiments can make only vague and approximate inferences. This situation is unfortunate because experimental data represent an expenditure of both time and money.

This book presents the view that a major part of planning for experiments should be consideration of appropriate statistical analysis before any data are gathered. The role of statistically designed experiments is to make the analysis of data as efficient and informative as possible. Although methods for analyzing data are the foundation for constructing statistically designed experiments, this book does not give statistical analysis methods. There are many existing books that fill this role; therefore, covering this topic would only detract from the emphasis on statistically designed experiments.

A growing objective of many experiments is the optimization of conditions for making chemical measurements. Chapter 5 presents both steepest ascent and simplex optimization methods for easily running sequential experiments to find instrument settings that maximize or minimize chemical measurements. Chapter 6 presents response surface experiments to characterize the surface of chemical measurements for different levels of the experimental factors. A bibliography of optimization and response surface methods, as actually applied in 17 major fields of chemistry, is provided in Chapter 7 as a guide to current practice.

Only an elementary background in probability and statistics should be needed to use this book. Some basic matrix operations are introduced in Chapter 4 to avoid long and tedious algebraic equations. Other mathematical equations require only introductory college algebra and calculus courses. In the appendix, three computer programs, written in BASIC on an IBM-PC, are given to calculate Student's t-distribution, to test for normality, and to calculate half-normal plots. These programs may have to be modified to run on other types of computers.

vi

We would like to thank our many friends for their help and encouragement while writing this book. Special thanks go to Dr. Max D. Morris, who read the manuscript and made many helpful improvements, and to Mrs. Pauline S. Bayne for able preparation of the illustrations and patience with her husband.

Charles K. Bayne
Ira B. Rubin

TABLE OF CONTENTS

1

Experimental Design Basics

An experiment is viewed by chemists and statisticians in different lights, but both are concerned with confirming or disproving something doubtful, discovering unknown principles or effects, or testing some suggested theory. Chemists require several components to produce experimental results related to programmatic objectives. The main components are procedure descriptions (eg, analytical methods), analysts, apparatus, reagents, laboratories, and test materials. Obtaining these components may require cooperation of colleagues, literature searches, or additional funding. Once experimental results are produced, the data are statistically analyzed to see if they accomplish the objectives.

Statisticians concentrate on three experimental components: statements of objectives, statistical analysis, and experimental design.

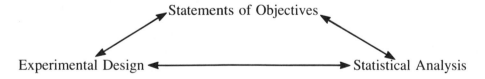

Each component depends on the others for its definition and structure. Conclusions from one experiment may redefine the statement of objectives for the next experiment. This new experiment may require a revised experimental design and statistical analysis method. Frequently, a series of small experiments are used to achieve the objectives, rather than one large experiment from which all conclusions are derived.

Statisticians' views and chemists' views of an experiment are complementary rather than contradictory. In fact, differences between chemists and statisticians usually arise because of problems with terminology and a failure to understand each other's discipline. Because each discipline can greatly benefit the other, statisticians must learn some chemistry, and chemists must learn some statistics. In this way, each will have a highly successful interaction. This book will attempt to translate some statistical ideas used in experimental design for chemists by using examples of experiments from the chemistry literature.

The importance of interaction between chemists and statisticians is emphasized by the introduction, in 1974 (Kowalski 1978), of the word *chemometrics* to define applications of mathematical and statistical methods to chemical measurements. Journals with papers related to chemometrics are *Analytical Chemistry*, *Analytica Chimica Acta*, *Analytical Letters*, *Chemometrics Section Newsletter*, *Computers in Chemistry*, *Computer Techniques and Optimization*, *Industrial and Engineering Chemistry*, *Journal of Chemical Information and Computer Science*, and *Technometrics*.

1.1   Preparation for Experiments

Rushing into a laboratory to immediately test a new method or to compare a new apparatus with older equipment may be exciting, but these forays may also be disappointing. Because nothing is proved scientifically, unfair comparisons are made, wrong parameters are estimated, concentration ranges are too narrow, temperature effects are not controlled, and so on. Frequently, data are generated with few results, while time is lost for valid scientific investigation. The hard part of experimental design is to realize that jumping into a laboratory to start experiments is the wrong course of action.

The first step in designing an experiment is to state the objectives exactly. This is simple advice, yet many studies fail to define their objectives and, as a result, experimental data neither confirm nor disprove the experimenter's expectations. Defining objectives helps to keep track of the experiment's goals in the midst of all its details. A statement of objectives is most important for studies lasting a long period of time, during which personnel or subcontractors may change. In reports on lengthy experimental studies, the objectives frequently are stated in the initial chapter, but by the concluding chapter these goals are nowhere to be seen. Results should always be evaluated in terms of the objectives.

Four important objectives are (1) comparing treatment effects, (2) estimating parameters, (3) deriving prediction equations, and (4) optimizing operating conditions. A *treatment* (Kendall and Buckland 1982) in an experiment is a stimulus that is applied to observe the effect on the experimental situation. In practice, treatment may refer to a physical substance, a procedure, or anything that is capable of controlled application. Comparison of treatment effects may be used, for example, to identify which of several treatments produces a maximum process yield. These comparisons are expressed in statistical terms as significance statements, such as "treatment A is significantly larger than treatment B at the 5% significance level." Statisticians may determine that there are significant differences among treatments. However, chemists must use their expertise to decide if the magnitude of the differences is relevant in practice.

Parameters are constants in mathematical or statistical models used to represent experimental responses. Parameters are usually unknown and must be estimated by a function of observed experimental values called an *estimator*. There are, in general, many possible estimators for any parameter. For example, both the average and median are estimators of mean response. Choosing among estimators is based on the expected values and standard deviations of the estimators. The expected value of an estimator should be equal to the parameter being estimated. Chemists usually refer to the difference between an estimator's expected value and the true parameter value as *accuracy*. And estimator's standard deviation is known as its *precision*. The numerical value of an estimator is an *estimate* of the parameter. How well an unknown parameter can be estimated is expressed in a confidence interval for the estimator. A *confidence interval* represents an interval in which the parameter value would occur with a specified probability if the experiment were repeated many times under exactly the same conditions. The construction of a parameter's confidence interval is usually based on the distributional properties of its estimator.

A *prediction equation* is a mathematical formula (usually a polynomial) that approximates experimental responses. Parameters of the approximating function (ie, coefficients) are frequently derived using least squares methodology. Chemists use prediction equations for two purposes: (1) to predict future values, and (2) to represent expected values of experimental responses. Derivation of calibration curves is an example of the first use. An example of the second use is a *response surface*, which is formed by graphing the prediction equation as a function of explanatory variables. A response-surface experiment is used to approximate this surface at values of the explanatory variables, especially where the response is a maximum or a minimum.

Optimization of laboratory procedures has frequently relied on the old vary-one-factor-at-a-time method. This method has been shown to be inefficient, and in many situations it can lead to incorrect results. *Steepest ascent* and *simplex optimization* are two experimental design methods being adopted by chemists to efficiently search for optimum conditions. Steepest ascent uses results of an initial experiment to calculate the gradient or direction of movement towards the optimum. Subsequent experiments are then performed along this path of steepest ascent. Simplex optimization requires very simple computation. Responses from an initial experiment need only to be ranked from worst to best. To move towards the optimum, one forms a new simplex by discarding the worst experimental value and replacing it with a new value in the opposite direction.

These four objectives—treatment comparisons, estimations, predictions, and optimizations—are interrelated. For example, once a treatment comparison has been made, parameters representing treatment effects can be estimated, and confidence intervals can be calculated about these parameters. Prediction equations can be used to estimate the expected value of a response at fixed values of the explanatory variables. Prediction equations may also be used to approximate a response surface to estimate the optimum value of an experimental response.

The objectives are accomplished by a statistical analysis of experimental data. Therefore, the objectives must be formulated either in terms of a hypothesis that can be tested or in terms of mathematical functions that can be estimated. The objective of showing that treatment A is better than treatment B may be translated to "the expected response from treatment A is significantly larger (at an agreed-on significance level) than the expected response from treatment B." Note that expected responses are unknown parameters that must be estimated. In statistical notation the *null hypothesis* H_o: $\mu_A = \mu_B$ is tested against the *alternative hypothesis* H_1: $\mu_A > \mu_B$ where μ_A and μ_B represent the expected response of treatment A and treatment B, respectively. Rejecting the null hypothesis by statistical analysis would indicate that treatment A is better than treatment B. Formulation of objectives into statistical terms gives an exact meaning to the phrase "treatment A is better than treatment B." If the real objective is to show that treatment A is different from treatment B, then the null hypothesis can be expressed as H_o: $\mu_A = \mu_B$, and the alternative hypothesis can be expressed as H_1: $\mu_A \neq \mu_B$. The exact meaning of the objectives should always be stated so that results are clear to any reader.

Statements of objectives should also include the scope of the final conclusions. For example, "treatment A is better than treatment B for polyaromatic hydrocarbons in the concentration range of 10–15 mg/ml." Defining the scope limits experimental conclusions to a set of conditions, but, in reality, experiments already have limits due to restrictions of the time and money. Therefore, it is important to select those conditions that have the most value for the objectives. Extrapolating conclusions of an experiment beyond its scope (frequently done with calibration curves and fitted equations) can lead to erroneous conclusions.

After defining the objectives, chemists need to think about the statistical analysis even before data are collected. Types of statistical statements that will be made (significant differences, confidence intervals, precision and accuracy statements) must be considered. Formulating such statements will force chemists to consider whether appropriate statistical methods exist to meet the objectives. If not, the objectives may have to be reformulated.

The majority of statistical analyses are done by representing an experimental response by its expected value plus an error term. The expected value is formulated as a mathematical function in terms of the explanatory variables.

$$\text{response} = \text{expected value} + \text{error term}$$

The set of error terms from all experimental runs is assumed to have a probability distribution. Assumptions about this probability distribution are used to derive statistical tests for stated hypotheses or to estimate parameters in the mathematical function. The most common assumptions about distributional properties for error terms are the following: they are independent, they have identical distributions with zero mean and identical variances, and each has a normal probability distribution.

Three important aspects of statistical analysis must be considered: (1) What is the proper response to answer the objectives? (2) Is the mathematical function used to represent the expected response reasonable? (3) Are the assumed distributional properties of the error terms satisfied? Answers to these three questions depend primarily on the experimental design and on the measurement chosen to represent the experimental response. Frequently, the experimental response is an algebraic function of basic measurements formed by addition, subtraction, multiplication, or division. Because distributional properties of these algebraic forms can be complex, it may be difficult or impossible to derive statistical tests or estimation properties. For example, for two measurements, each having a normal distribution, the ratio of the measurements does not have a theoretical expected value or variance, and the product of the measurements has a distribution that is an infinite sum of Bessel functions. To avoid difficult statistical analysis problems, whenever possible one should choose the experimental response to be either a basic measurement or a linear combination of basic measurements.

The expected value of the response is expressed as a mathematical function in terms of explanatory variables. This mathematical function is sometimes derived from theoretical considerations of the response. However, in the majority of experiments, either the theory does not exist, or it is so complex that the mathematical form cannot be written in a closed form. Therefore, the true mathematical function is approximated. The basic approximating function is a low-order polynomial. Motivation for using polynomials as approximating functions is the Taylor series theorem, which states that, under certain differential and continuity restrictions, values of a function can be approximated as closely as needed by a polynomial of sufficient degree. One purpose of an experimental design is to provide information to evaluate whether a low-order polynomial, rather than a higher-order polynomial, is sufficient to represent the true mathematical function.

After considering both the objectives and statistical analysis, the experimental design can be constructed. For a statistician, constructing an *experimental design* means designating the number of experimental units and the order of applying treatments to experimental units. The number of experimental units is determined by considerations such as the number and type of explanatory variables, the number of parameters to be estimated, the desired probability of detecting differences, time, money, and so on. The order of allocating treatments to experimental units is established by randomization to insure unbiased estimates of treatment effects and their reproducibility. An *experimental unit* is that entity that is allocated a treatment independently of other such entities. It may contain several *observational units*. For example, a sample is spiked with a fixed level of radioactive carbon and then split into three subsamples. The sample is the experimental unit, and the subsamples are the observational units. The key phrase in the experimental unit definition is "independently of other such entities." Distinction between experimental and observational units is important for determining measurement precision. This distinction will be explored in later chapters. An *experimental run* is the application of a treatment to an experimental unit and the resulting response measurements.

Experimental designs can be constructed to minimize the effect of extraneous variables, such as ambient temperature, humidity, or air pressure on the response. The definition of an extraneous variable is similar to the definition of a weed—all those variables that we would like to ignore in a particular experiment but cannot, due to their known or assumed effect on responses. For the next experiment, those weeds may turn out to be the flowers. The second major purpose for designing an experiment is to guarantee that model assumptions are correct or to provide a method of checking assumptions. Two major assumptions are concerned with adequacy of the approximating model and distribution properties of the error term. Two methods used to guarantee and to check these assumptions are replicating experimental units and randomizing the order in which experimental treatments are applied.

1.2   Checklist for Experimentation

The following list of questions will assist in the preparation and design of an experiment.

Statement of Objectives

1. Why is the work being done? What questions should the experiment answer?
2. What are the consequences of a failure to find an effect or to claim one when it does not exist?
3. How are the questions represented with mathematical or statistical models?
4. What is the best response to measure to meet the experimental objectives?
5. How will the data be collected and stored? Should a computer data-collection system be implemented?
6. What are the time schedule and allowable costs?

Statistical Analysis

1. What type of statistical statements will be made about the experimental results?
2. What type of statistical model is assumed to analyze the data?
3. What are the assumptions of the statistical model, and how can they be examined using the experimental data?
4. What statistical methodology is used to analyze the statistical model?

Experimental Design

1. What are the explanatory variables, extraneous variables, and response variables? What are the ranges of the explanatory variables and the levels to be measured?

2. Which explanatory variables are considered most important? Do any of the variables interact?
3. How can the extraneous or disturbing variables be controlled or minimized?
4. Is there any information about the precision of the measured response? Is there any similar experiment in the literature? How many experimental runs can be made? Will there be any replication?
5. Can a standard experimental design be used, or must a design be tailored to this experiment?

2

Preliminary Planning

During preliminary planning of an experiment, decisions must be made about experimental objectives, statistical analysis, and experimental design. These decisions include specific details about which variables will be primary, background, or fixed; the need for randomization; what the experimental units are; how many experimental units are needed; how multiple responses are to be analyzed; and proper methods for recording data. Examples illustrating the thought process and methods for making these decisions are given in the following sections.

2.1 Experimental Objectives

Four important objectives for doing experiments were listed in Chapter 1: (1) comparison of treatment effects, (2) estimation of parameters, (3) derivation of prediction equations, and (4) optimization of operating conditions. These four objectives will be examined in this section by using hypothetical examples typical of problems that chemists encounter.

2.1.1 Comparison of Treatment Effects

Objective
Test the hypothesis that high-pressure liquid chromatography (HPLC) gives the same naphthalene values as standard liquid chromatography for analyzing synfuels from coal.

Statistical Model
Let $Y_{i,j}$ = the naphthalene value on the *j*th experimental unit analyzed

9

by the ith treatment. Treatment $i = 1$ will represent the standard liquid chromatographic analysis, and treatment $i = 2$ will represent the HPLC analysis. The index $j = 1, 2, \ldots, N$ represents the experimental unit or the synfuel sample to be analyzed by the two treatments.

The probability distribution for each observation is assumed to be a normal distribution with each observation independent of the other observations. The expected value of each observation is $E(Y_{ij}) = \mu_{ij}$, with a variance of σ^2. These statements are summarized by $Y_{ij} \sim$ i.i.d. $N(\mu_{ij}, \sigma^2)$, where i.i.d. stands for "independent, identically distributed." Greek letters μ and σ^2 represent the expected value and the variance of the true population. Their values are usually unknown and must be estimated from a sample from the true populations. Estimates of the mean and variance from sampled values are denoted by $\bar{Y}$ (Y bar) and S^2, respectively.

Experimental Design

Ten different synfuel samples are available for testing. Two experimental units are made from each sample, and the paired experimental units are randomly ordered. This randomization can be done by numbering the paired experimental units from one to ten and selecting ten numbers from a hat to determine the order of analyzing the paired experimental units. The treatment used to analyze the first experimental unit in each pair is decided by a coin toss.

TABLE 2.1. Paired t-test for comparing HPLC and standard liquid chromatography methods for analyzing naphthalene. Each method was used on ten experimental units

	Naphthalene Values									
HPLC	12.1	10.9	13.1	14.5	9.6	11.2	9.8	13.7	12.0	9.1
LC	14.7	14.0	12.9	16.2	10.2	12.4	12.0	14.8	11.8	9.7
Difference	−2.6	−3.1	0.2	−1.7	−0.6	−1.2	−2.2	−1.1	0.2	−0.6

Mean of the difference $= \bar{D} = -1.27$
Standard deviation of the difference $= S_D = 1.13$
t-statistic $= \bar{D}/(S_D / \sqrt{10}) = -3.55$ with nine degrees of freedom

Statistical Analysis

The null hypotheses is H_o: $\mu_{1j} = \mu_{2j}$; the alternative hypothesis is H_1: $\mu_{1j} \neq \mu_{2j}$. The data, shown in Table 2.1, are analyzed by a paired t-test (Snedecor and Cochran 1967) at the 5% significance level. This level means that there is a 1 in 20 chance of accepting the alternative hypothesis when, in fact, the null hypothesis is true.

For a two-sided hypothesis test, the absolute value of the t-statistic is compared to the 2.5% percentile point of the *t*-distribution with nine degrees of freedom. Because $|t| > 2.262$, the *t*-statistic would be considered larger than expected under the null hypothesis, and, therefore, the naphthalene values from the two methods of analyses would be considered significantly different at the 5% significance level.

2.1.2 Estimation of Parameters

Objective
Given N_0 grams of a radioactive gas, estimate its half-life.

Statistical Model
Because radioactive decay is a Poisson process, the time between decays (interarrival time, T) has a gamma distribution with parameter A (Parzen 1962):

$$g(T) = A \exp(-AT), \quad 0 < T < \infty$$

The parameter A is the decay rate (min^{-1}); if it can be estimated, the half-life can be determined by (Sheehan 1961) the formula:

$$t_{1/2} = \frac{N_0 \ln{(2)}}{A}$$

Experimental Design
Suppose the gram molecular weight of the radioactive gas under standard conditions can be determined, and by Avogadro's law we calculate that there are $N_0 = 6.024 \times 10^8$ atoms. Next, decays from the radioactive gas are detected, and the interarrival times are recorded for ten decays, with the following results:

Statistical Analysis

Decays	1	2	3	4	5	6	7	8	9	10
T (sec)	1.00	0.34	0.51	0.16	0.45	0.51	0.29	0.29	0.59	1.18

The probability of getting the results $T_1, \ldots, T_n$ depends on the product of the density functions, $g(T_1)g(T_2) \ldots g(T_n)$. A method of estimating

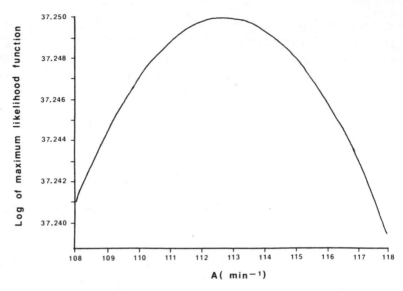

Figure 2.1.   Logarithm of the maximum likelihood function versus values of parameter A.

parameter A in the density functions is to find the value of A that yields the maximum probability of observing the sample results. This estimate of A is called the *maximum likelihood* estimate. An equivalent procedure is to find an estimate of A that maximizes the logarithm of the product of density functions, denoted by ML(A).

$$ML(A) = n \ln(A) - A \sum_{i=1}^{n} T_i$$

$$= 10 \ln (A) - 8.87 \times 10^{-2} A.$$

In Figure 2.1, ML(A) is plotted against values of A ranging from 108 to 118 min^{-1}. We see that a maximum is reached when $A = 112.5$ min^{-1}. The exact value of the maximum likelihood function is found by differentiating ML(A) and equating the result to zero. This value is $A = 10/8.87 \times 10^{-2} = 112.740$ min^{-1}. Therefore, the estimated half-life is

$$t_{1/2} = \frac{N_0 \ln (2)}{A}$$

$$= \frac{(6.024 \times 10^8) (0.693)}{112.740}$$

$$= 3.703 \times 10^6 \text{ min or } 7.05 \text{ years.}$$

2.1.3 Prediction Equations

Objective
Derive a mathematical equation to predict the acid to carbon concentration when sugar is added to nitric acid over a 24-hour (1440 min) time period at a high temperature.

Statistical Model
Let Y_i represent the logarithm of observed acid to carbon concentration at time t_i. The expected value of Y_i is assumed to have the functional form $E(Y_i) = A + Bt_i + Ct_i^2$, and the distribution of the observations is $Y_i \sim$ i.i.d. $N[E(Y_i), \sigma^2]$. An equivalent model is written in terms of the observed values as $Y_i = E(Y_i) + e_i$, where the e_i's are the random error terms with a distribution of $e_i \sim$ i.i.d. $N(0, \sigma^2)$.

Experimental Design
For each observed value, add a measured amount of sugar (ie, experimental unit) to nitric acid and let the reaction run for time t (min) (ie, treatment); quench the reaction and measure the acid and carbon concentrations. The time to quench will be measured in equally spaced intervals to provide information on the curvature of the quadratic function.

In addition, we would like to test whether our model has a lack of fit; therefore, each experimental run will be duplicated. Because the work must be completed in six weeks or 30 working days, the number of experimental runs will be determined by how much work can be done in that time frame. Suppose an experimental run was made for quench times that were multiples of 4-hour intervals and that the time required to run a zero quench time is negligible.

Time to quench (hrs)	0	4	8	12	16	20	24
Days required	0	1	1	1.5	2	2.5	3

We see that seven duplicate runs require 22 days. If 10% of our time (three days) is used to analyze the data and write up the results, we have a margin of five days for setting up the apparatus and allowing for mistakes. When deciding the number of experimental units, all facets of the experiment must be considered: setup time, experimental time, time to analyze the data and write up the results.

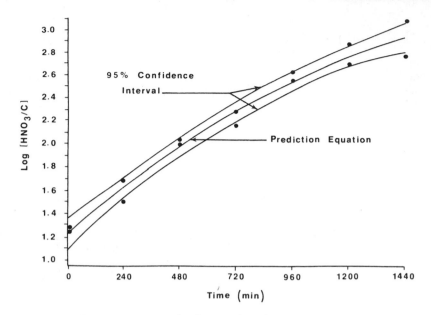

Figure 2.2. Prediction equation for log(acid/carbon) versus time.

Before starting the experiment, the order of running experimental units should be randomized. Frequently, experimenters think that when duplicate runs are made they have to be run in sequence. However, if we recall the definition of an experimental unit, the units must be run independent of each other to prevent any correlation from one experimental run to the next. The experimental units are individual sugar samples; treatments (eg, quench time) are randomly assigned, to the experimental units. Each quench time is assigned two numbers from 1 to 14 (because of duplicate runs), and these may be assigned sugar samples by selecting numbers randomly from a hat.

A random order would be to run the time intervals in the following sequence:

$$4,24,0,24,20,20,12,0,4,16,8,16,8,12.$$

By simply examining the sequence of numbers, we are unable to determine if they are randomly selected. The selection process is what determines that a sequence of numbers is random.

Statistical Analysis
The data are plotted in Figure 2.2, and a least squares fit of a quadratic equation is given by

$$\hat{Y}_i = 1.26 + 1.67 \times 10^{-3}\, t_i + 4.46 \times 10^{-7}\, t_i{}^2,$$

where $\hat{Y}_i$ (eg, Y "hat") is the estimated expected response at t_i .
   The sum of squares of the error term,

$$\text{SSE} = \sum_{i=1}^{14} (Y_i - \hat{Y}_i)^2,$$

can be partitioned into two parts. The first part represents the sum of squares
due to duplicate observations; the second represents the sum of squares due
to the inadequate fit of the quadratic model. The first part is called *pure
error*,

$$\text{SSPE} = \sum_{i=1}^{7} \sum_{r=1}^{2} (Y_{ir} - \overline{Y}_i)^2,$$

where Y_{ir} is the *r*th measurement (eg, *r*th replicate) made at t_i , and $\overline{Y}_i$ is the
average of the two replicates at t_i . The second part is called *error due to lack
of fit*,

$$\text{SSLOF} = \text{SSE} - \text{SSPE},$$

and can be compared to the pure error by an *F*-test to determine the
adequacy of the fitted model:

$$F\,(k-p, n-k) = \frac{\text{SSLOF}/k-p}{\text{SSPE}/n-k.}$$

The *degrees of freedom* for the *F*-test in this example are $k-p = 4$ and
$n-k = 7$, because there are $n = 14$ experimental units, $k = 7$ experimental
treatments, and $p = 3$ estimated paramenters. If the *F*-test is not significant
at the 5% level, then $\text{SSE}/(n-p)$ can be used as an estimate of the variance
of the error term. If the *F*-test is significant, we must investigate the
adequacy of our model. This lack of fit may be due to outlier values,
incorrect distributional assumptions, or an incorrect model. Each of these
can be examined by residual plots (Draper and Smith 1981).
   Confidence intervals can be placed on $\hat{Y}_i$, as illustrated in Figure 2.2,
using standard regression-analysis methods. Remember that confidence
intervals on $\hat{Y}_i$ are for expected values rather than a single observation that
may be observed in the future.

2.1.4   Response Surface

Objective

Find the optimum air pressure and bead heating current for the rubidium bead type of nitrogen–phosphorus gas chromatographic detector to detect the maximum peak areas of carbazole and quinoline.

Statistical Model

Let C and Q represent observed carbazole and quinoline peak areas in $cm^2/\mu g$. Assume the observations can be modeled as second-order functions in terms of air pressure, P, and bead heating current dial setting, B:

$$C = A_0 + A_1 P + A_2 B + A_3 P{*}B + A_4 P^2 + A_5 B^2 + e,$$
$$Q = B_0 + B_1 P + B_2 B + B_3 P{*}B + B_4 P^2 + B_5 B^2 + f,$$

where the random error terms are $e \sim$ i.i.d. $N(0,\sigma_1^2)$ and $f \sim$ i.i.d. $N(0,\sigma_2^2)$. We want to find values of P and B for which C and Q have maximum values.

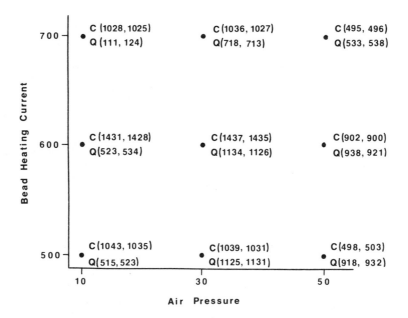

Figure 2.3.   Three-level factorial design for the carbazole–quinoline response surface example. Replicate observed values are C for carbazole and Q for quinoline.

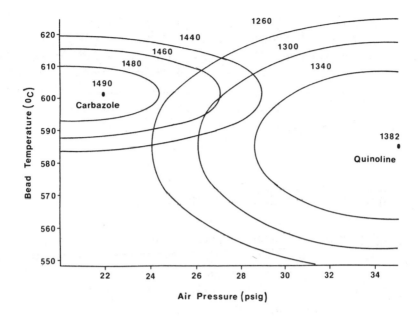

Figure 2.4. Contours of the response surface equations for carbazole and quinoline peak areas $(cm^2/\mu g)$.

Experimental Design

The experimental units are displayed in Figure 2.3 in a three-level factorial design (see Chapter 4). For this experiment the two factors are bead temperature dial setting, B, and air pressure, P. The three levels for each factor are $B = 500°C$, $600°C$, and $700°C$, and $P = 10$, 30, and 50 psig. We note from Figure 2.3 that all combinations of the levels from each factor are represented in this design. These combinations should be performed in random order rather than fixing each level of B and doing all three levels of P. Replicating each experimental unit provides information on the model's lack of fit and the error assumptions.

Statistical Analysis

The estimated prediction equations for C and Q are found by using least squares in the same manner as Example 2.2.3. The estimated prediction curves are

$$E(C) = -13180 + 27P + 48B - 0.67P^2 - 0.04B^2,$$
$$E(Q) = -6284 + 70P + 22.7B - 1.00P^2 - 0.02B^2.$$

Coefficients A_3 and B_3 were not significantly different than zero and were eliminated from the models. Using differential calculus, we determine that estimated maxima are 1490 for carbazole at $(P=21.8, B=600)$ and 1382 for quinoline at $(P=35, B=567.5)$. Because we have to compromise between two maxima, contours of the two response surfaces are plotted in Figure 2.4. We see from these contours that a reasonable operating value is $(P=29, B=590)$, where the values of the two responses are $C = 1438$ and $Q = 1328$.

2.2   Experimental Variables

An important part of planning an experiment is to identify *explanatory variables*—those variables that affect the measured response. The amount of time spent considering the number and the value of explanatory variables will pay big dividends. The best method is to make a list of every possible variable that may affect the response. Some may seem unlikely, but put them down anyway. Now divide the list into primary variables and blocking variables.

The *primary variables* are those variables that we suspect may have the largest effect on the response. A subset of the primary variables will represent the explanatory variables in the experiment. Primary variables can be classified either as quantitative or qualitative. *Qualitative variables* have values that cannot be ordered. For example, two primary variables for determining the best method to analyze uranium are the analytical method and the analyst. The analytical method may have two values, Davies–Gray titration and x-ray fluorescence; and the analyst may have three values of technicians, 1, 2, and 3. The values of the two primary variables cannot be put on a numerical scale or be ordered from highest to lowest. *Quantitative variables* have numerical values that can be ordered. In the example given in the subsection entitled ''Response Surface,'' the two primary variables of air pressure and bead heating current are both quantitative variables because both variables can be measured on a continuous numerical scale. Frequently, the primary variable is represented by a control switch on an instrument. If the switch has discrete settings that represent a change in conditions such as orange, red, green, or blue filters, the primary variable would be considered qualitative because a numerical order cannot be assigned to colored filters. If the switch has either discrete or continuous settings that represent an increase in voltage, temperature, air flow, and so on, the primary variable would be considered a quantitative variable even though we may not know the exact numerical values the switch settings represent.

Primary variables are divided into two categories. Those under investigation in an experiment are *factors*, and those having a fixed value during an experiment are *fixed variables*. For example, two primary variables in a conductivity experiment are pH and temperature. The objective is to study the conductivity response with changing values of pH at room temperature. In this case, pH is a factor, and temperature, which is fixed at room temperature, is a fixed variable.

The values of a factor are called its *levels*. The levels of a qualitative factor such as analytical methods could be many possible methods; however, we select only those methods that are relevant to the experiment. This restriction on the number of levels defines the range of qualitative factors. For quantitative factors, the range is chosen by specifying their minimum and maximum values to be assigned during the experiment. The factor levels are at least the minimum and maximum values and may also include other values in the range.

The levels of a factor may also represent groups of values. For a factor such as moisture of a sample, with levels assigned as dry, damp, and wet, the levels would be defined by a range or group of values. For example, dry may be defined as values of less than 25% humidity. For an experiment with a single factor, each level represents a treatment. An experiment with two factors, A and B, each with two levels, will have four treatments defined by all possible combinations of their levels: (a_1,b_1), (a_1,b_2), (a_2,b_1), and (a_2,b_2), with the a's and b's representing the levels of factors A and B, respectively. In a full-factorial experiment, all possible treatments are run; however, there are many experiments that select only some of the treatments. By defining both the range of the factors and the values of the fixed variables, the scope of the experiment is specified.

When selecting factors for an experiment, it is important to consider their units of measurement. Suppose two factors in an experiment on conductivity are nitric acid and temperature. If the nitric acid concentration is measured in terms of weight per unit volume, the amount of nitric acid will change values as the levels of termperature increase due to the effect of temperature on the volume of liquid. A more appropriate concentration unit for the nitric acid here would be a weight per unit weight. Factors expressed in percent units are also likely to be affected by other factors in the experiment.

Blocking variables are those variables that may affect the response but are neither factors nor fixed variables, such as sample materials, apparatus, analysts, and reagents. A common blocking variable in experiments is time, usually in units of a work day. Here we would block or group all treatments to be run in one working day. Because of changes in humidity, barometric pressure, and temperature, a response from the same treatment may vary from day to day more than the variation expected due to chance. We cannot control the day effect and may not be interested in the value of its effect. However, if all the treatments are run each day, the effect due to different days can be isolated by using the method of analysis of variance.

Experimental units should be selected to be identical, so that different responses are due only to different treatments. Frequently, the most homogeneous experimental units are grouped into blocks, and each block is represented by a level of the blocking variable. Blocking variables generally increase the sensitivity of detecting significant treatment effects. If there are more treatments than experimental units grouped in one level of a blocking variable, the experiment should be designed so that block effects are orthogonal to (ie, independent of) treatment effects. These orthogonal designs allow the effect due to blocking variables to be removed by statistical analysis.

Chemists should be aware that responses for a factor may depend on the level of a blocking variable. Suppose we would like to measure cresol in coal samples by two methods, A and B. We have ten coal samples, with each sample representing a level of the coal blocking variable. Each level may contain a different concentration of cresol. From each level, two experimental units are prepared, with method A applied to one unit and method B applied to the other unit. If measurements from method A are always greater than measurements from method B for levels with cresol concentrations of 0.5 mg/mg or greater and the reverse is true for levels with cresol concentrations less than 0.5 mg/mg, then there is an interaction effect between the method factor and the coal blocking variable.

2.3   Experimental Error and Randomization

After defining factors, fixed variables, and blocking variables in an experiment, we realize that there may be other sources of variation that affect the response. These sources may include variability due to (1) measuring or recording the response, (2) inability to reproduce the treatments, (3) inability to reproduce homogeneous experimental units, (4) interaction of treatments with experimental units, and (5) explanatory variables that are unknown and beyond the control of the experimenter. All these sources are represented by one term called *experimental error*. The effect of experimental error is balanced out over treatments by the process of randomization.

Randomization is a method of assigning treatments to experimental units so that every possible arrangement of treatments on experimental units has an equal probability of occurring. Emphasis here is on randomization methods and not on the resulting assignments. For example, we cannot look at a set of numbers and determine whether randomization has been accomplished; instead we must examine the method of producing the numbers. Randomization is not the same as making haphazard assignments of treatments to experimental units, and it cannot overcome poor experimental technique. Methods of randomization require some device of chance, such as pulling numbers from a hat, flipping a coin, random number tables (Abramowitz and Stegun 1972), or computer-generated random numbers. Suppose eight experimental units are to receive four replicates of two treatments. Assign numbers 1 to 8 to the experimental units and select eight random numbers from a random number table. For example, the following two-digit random numbers were selected:

$$53 \ 47 \ 98 \ 11 \ 15 \ 03 \ 61 \ 22$$
$$6 \quad 5 \quad 8 \quad 2 \quad 3 \quad 1 \quad 7 \quad 4$$

Ranks are then assigned to the random numbers: Thus, random number 03 is the smallest and receives rank number 1, random number 11 is second and receives rank number 2, and so on. These ranks are considered to be a random permutation of the numbers 1 to 8. Therefore, treatment 1 would be assigned to experimental units labeled 6, 5, 8, and 2, and treatment 2 would be assigned to experimental units labeled 3, 1, 7, and 4. This randomization method may be extended to experimental designs in which some factors can be varied more easily than others. Examples of these types of designs are given in Chapter 3.

The purpose of randomization is (1) to protect against unsuspected sources of bias in order to get unbiased estimates of both treatment effects and experimental error variance; and (2) to obtain independent responses. Because experiments are affected by sources that either are overlooked or not identified, randomization is needed to impartially distribute these experimental errors over treatments.

Independent responses mean that the probability of observing any specific response value is not influenced by previously observed values. Frequently, lab technicians can recognize quality control samples and have a preconceived notion of their concentrations, based on past quality control samples. Responses observed sequentially on a treatment may not be independent because personal biases cause large values to follow large values or small values to follow small values. Randomization is a method to overcome preconceived notions or personal biases, so that responses can be assumed to be independent. The assumption of independent responses or independent experimental errors is crucial to many statistical analyses. The independent assumption allows us to calculate theoretical probability distributions for test statistics. Using these distributions, we can judge whether the experimental observations support the null hypothesis. The process of randomization in an experiment gives a theoretical basis for the independent assumption.

2.4   Experimental Units

Experimental units are those units that are allocated treatments independently of other units. Experimental units are usually easy to identify in agriculture experiments, for which many early principles of experimental design were derived. They are usually plots of grounds, pots of soil, groups of animals, and so on. In chemistry, however, experimental units are not always easy to identify, because most chemistry experiments consist of a series of operations, such as preparation of standard solutions, preparation of samples to be analyzed, pipetting, weighing, and titration. Frequently, only repeated analysis of the last step in a procedure is used to represent replicated experimental units. Therefore, only the precision of the last step can be estimated, but no information is gained about previous steps. Suppose we wish to examine an analytical method that consists of measuring 5 μg/ml benzo(a)pyrene (BAP) by gas chromatography (GC). We can make up a standard solution of 5 μg/ml of BAP and measure five aliquots by GC. From these measurements we could estimate the precision of repeated GC measurements. However, if we made five standard solutions and measured one aliquot from each solution, we could estimate the precision of the analytical method itself. Frequently, preparation of samples and standard solutions are major contributors to the error for an analytical method. For the first experiment, the treatment consists of only GC measurements; in the second, the treatment consists of preparation of a standard followed by a GC measurement. In the first case, experimental units are aliquots from a single standard solution; in the second case, experimental units are aliquots from different standard solutions. Determination of experimental units depends on the experimental objectives. The conditions investigated should be as representative as possible of the conditions on which the inferences will be made.

If the objective in the BAP experiment were to measure precision of the analytical method and we followed the first scenario of repeated GC measurements, our estimated precision would be too small. Precision estimated from repeated measurements on the same experimental unit is called the *observational unit error*, which measures the failure of the responses on any experimental unit to be precisely alike. This error is of little value for assessing treatment differences or determining lack of fit of a model if other sources of error are present. The second method is preferable because it measures both variation due to sample preparation and variation due to GC measurements.

2.5   Number of Experimental Units

When treatments appear more than once in an experiment, they are said to be replicated. The purpose of replication is to
1. Determine experimental error for treatment comparisons
2. Improve precision of estimated parameters
3. Check adequacy of the fitted model
4. Check variance assumptions

For an unreplicated experiment there is no direct estimate of experimental error. In this case, estimates of experimental error are based on prior knowledge of the experiment or the assumption that the variance associated with multifactor interaction is equivalent to experimental error. This is the usual course of action for experiments with many factors, or when limitations on the number of experimental units are necessary. Replicate measurements should be made whenever possible, because estimated precision of experimental error from replicate measurements is valid regardless of the assumed model.

The number of replicate measurements depends on the number of experimental units required, as determined by resources available and the desired objectives. The first step is to determine the total number of possible experimental units that may be needed. In a full-factorial experiment with K factors, the total number of experimental units equals the number of treatment combinations times replicates, or

(levels for factor 1)$\times \cdots \times$(levels for factor K)$\times$replicates.

For example a two-factor experiment with factor A having four levels and factor B having five levels, with each experimental unit replicated twice, has a total of $4 \times 5 \times 2 = 40$ experimental units. If each experimental unit takes one hour to run, five working days are required to do the experiment. If we allow time for sample preparation, equipment failures, data analysis, and reporting results, this project could require three working weeks to complete. The actual experimental runs may be only a small part of the total time required. If a cost can be associated with each experimental unit in terms of time or price, this cost can be used to calculate the maximum number of experimental units that can be run while staying within the budget. It is important to consider all aspects of the experiment, from sample preparation to report writing, to determine the possible number of experimental units that can be run. If all possible treatment combinations cannot be run, several options are available:

1. Reduce the number of factors by assigning fixed values to some factors. This action reduces the scope of the experiment; however, conclusions cannot be as general.
2. Reduce the number of levels used for each factor. This choice may reduce the order of the effects that can be determined for each factor. For example, linear effects can be estimated, but quadratic effects cannot.
3. Reduce the number of experimental runs by using experimental designs such as fractional factorial designs. Although these designs reduce the order of the effects that can be estimated, they use a small number of experimental runs to investigate many factors.

After considering available resources, one can use additional criteria to determine the number of experimental units. Such criteria include the desired precision of the estimated response and the probability of detecting a significant difference between two treatments. Sample sizes based on these two criteria will be described in the next two sections.

2.5.1 Sample Size Based on Precision

Suppose we wish to estimate the expected value, μ, of a response by its average value, $\bar{Y}$. We would like the absolute difference between the estimated and expected values to be less than L with a probability of $1 - \alpha$; that is, $\Pr(|\bar{Y} - \mu| < L) = 1 - \alpha$. For this problem we must specify the following:

1. The limit of error, L
2. The probability level, $1 - \alpha$
3. The standard deviation, σ, or an estimate of its value

TABLE 2.2. Common normal percentile points $Pr(Z > Z_\alpha) = \alpha$

$Z_{0.20}$	$Z_{0.10}$	$Z_{0.05}$	$Z_{0.025}$	$Z_{0.01}$	$Z_{0.005}$
0.848	1.282	1.645	1.960	2.326	2.576

The number of experimental units, N, is approximated by

$$N = \left(\frac{Z_{\alpha/2}\sigma}{L}\right)^2 .$$

The common normal percentile values are given in Table 2.2. Note that the percentile point used is $Z_{\alpha/2}$ rather than Z_α, because the probability level is for two-sided limits on the expected value (ie, absolute value). The limit of error and standard deviation measurements must be in the same units. If L is given in absolute units, then σ must be in absolute units. Relative units may also be used.

For example, we would like to estimate the sulfur content in a fuel oil within ± 0.2 g/l. The probability that the true value lies within the error limits should be 0.90, and from past measurements we know that the standard deviation can be estimated as $S = 0.4$. The number of experimental units is

$$N = \left(1.645 \times \frac{0.4}{0.2}\right)^2 = 10.8,$$

or $N = 11$.

In another experiment we wish to estimate the amount of benzo(a)pyrene in coal within $\pm 5\%$, and the analytical method is 10% accurate. The probability desired is 95% for the true value to be in the limits of error.

$$N = \left(1.960 \times \frac{10}{5}\right)^2 = 15.4,$$

or $N = 16$.

2.5.2   Sample Size Based on Hypothesis Tests

The second method for approximating the number of experimental units is based on testing for a significant difference between expected values of two treatments. For each treatment, observations will be measured on N experimental units for a total of $2N$ experimental units. Treatment observa-

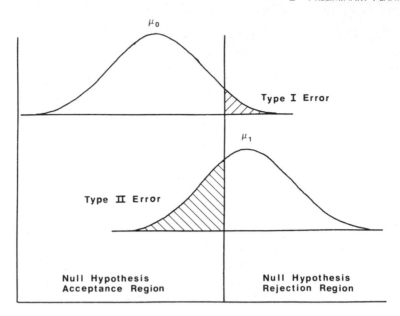

Figure 2.5.   One-sided hypothesis test illustrating type I and type II errors. Type I error is the probability of declaring a difference when there is no true difference. Type II error is the probability of declaring no difference when there is a true difference.

tions are assumed to be independent with a normal distribution. For this method we must decide

1. The standard deviation or an estimate of its value
2. The detectable difference between the expected values
3. The probability of declaring a difference when there is no true difference (type I error)
4. The probability of declaring no difference when there is a true difference (type II error)
5. The type of hypothesis to be tested
    One-sided test: H_0: $\mu_1 = \mu_2$ versus H_1: $\mu_1 < \mu_2$.
    Two-sided test: H_0: $\mu_1 = \mu_2$ versus H_1: $\mu_1 \neq \mu_2$.

In Figure 2.5, two probability distributions illustrate type I and type II errors for one-sided hypothesis tests.

Two normal populations used to represent observations for treatment 1 and treatment 2 will be denoted by $N(\mu_1, \sigma_1^2)$ and $N(\mu_2, \sigma_2^2)$, where the μ's represent expected values and the σ^2's represent variances. The normal percentiles for type I and type II errors are represented by Z_I and Z_{II}, respectively.

Case I. $\mu_2 = \mu_1 + C\,\sigma_1$ and $\sigma_2{}^2 = H\,\sigma_1{}^2$.

For this case the detectable difference between the two expected values is expressed in units of the standard deviation of the first population. The variance of the second population is represented by a multiple of the variance of the first population. This case corresponds to testing the null hypothesis H_0: $\mu_1 = \mu_2$ versus the alternative hypothesis H_1: $\mu_2 = \mu_1 + C\,\sigma_1$, or, equivalently, H_0: $C = 0$ versus H_1: $C > 0$. The approximate number of experimental units for specified type I and type II errors is

$$N = \frac{(\sqrt{2}\,Z_I + \sqrt{1+H}\,Z_{II})^2}{C^2}.$$

This approximation is the number of experimental units for one treatment; therefore, the total number of experimental units needed for the experiment is $2N$.

For example, uranium can be measured by dichromate titration and also by x-ray fluorescence. We would like to test whether the difference between the uranium measurements by the two methods is more than $C = 1.5$ standard deviations of the titration measurements. The variance of x-ray fluorescence measurements is considered to be twice the variance of titration measurements. If we want type I error to be 5% and type II error to be 10%, how many experimental units are needed for each method?

$$N = \frac{(1.414 \times 1.645 + 1.732 \times 1.282)^2}{2.25} = 9.2,$$

or $N = 10$.

Therefore, we would need 20 experimental units, 10 for each method. For a two-sided hypothesis test, we would use $Z_{0.025}$ for Z_I, which would result in $N = 12$ experimental units per treatment. In general, a two-sided hypothesis test with a desired type I error of α would use $Z_{\alpha/2}$ in the approximating formula, but Z_{II} would not be changed. Exact values of the standard deviations or variances for the two populations do not have to be known. However, the desired difference to be detected between treatment means must be evaluated in terms of multiples of the standard deviation (ie, value of C), and the relationship between the two variances (ie, value of H) must be determined. Frequently, chemists can approximate values of C and H better than they can the value of σ_1.

Another method may be used if chemists can express the mean of the second population as a multiple of the mean of the first population, and if they know the relative error of the first population.

Case II. $\mu_2 = K \mu_1$ and $\sigma_2^2 = H \sigma_1^2$.

For case II we wish to determine if the mean of the second population is K times the mean of the first population. If the relative error, $RE = \mu_1/\sigma_1$, is known for the first population, then the number of experimental units needed is approximated by

$$N = \frac{(RE)^2 (\sqrt{2} Z_I + \sqrt{1+H} Z_{II})^2}{(K-1)^2}.$$

Again, this is the number of experimental units per treatment for a total of $2N$ experimental units.

Suppose we wish to determine whether the value of manganese measured by the bismuthate method is 1.1 times larger than the value of manganese measured by the Volhard method. The percent relative error is 5% for the Volhard method, and the variance of the bismuthate method is twice as large as the Volhard method. The number of experimental units to test H_0: $\mu_1 = \mu_2$ versus H_1: $\mu_2 = 1.1\mu_1$ with 5% type I error and 10% type II error is

$$N = \frac{(.05)^2 (1.414 \times 1.645 + 1.732 \times 1.282)^2}{(0.1)^2} = 5.2,$$

or $N = 6$.

Therefore, 12 experimental units would be required to test the null hypothesis. Again, if a two-sided test is desired for a type I error of α, we would use $Z_{\alpha/2}$ rather than Z_α.

The methods for approximating the number of experimental units in this section should be used as guides, because we usually do not know the exact values of the parameters C, H, K, or RE. The best approach is to plot the number of experimental units versus a range of the parameter values and, from these graphs, to choose a number that is reasonable with available resources. These numbers are based on comparisons of two treatments, but often we wish to compare several treatments. The two-treatment comparison case can be considered an upper bound on the number of experimental units per treatment. Additional methods for sample size determinations are given by Odeh and Fox (1975).

2.6   Multiple Responses

During optimization of experimental conditions, the chemist frequently makes several response measurements. Usually some measurements are to be maximized, and others are to be minimized. For example, when optimizing treatment conditions to maximize the concentration of a substance, we would also like to minimize impurities. Two methods are advocated to deal with multiple response problems:

1. Superimposing contour diagrams
2. Constructing a desirability coefficient

Superimposing contour diagrams for each individual response is a simple idea. For each individual response, we derive a response surface equation (see Chapter 6) and superimpose contour plots on the same graph. From this graph we can usually identify the most desirable treatment conditions. Although these conditions may neither maximize nor minimize each individual response, they will represent a reasonable compromise among all the responses. This method was used to determine a reasonable operating condition in Figure 2.4 for the carbazole–quinoline response surface example.

Problems encountered in plotting contours are that the number of factors may be too large to visualize the contour plots in higher dimensions, and the metrics of the responses may not be compatible for plotting on the same graph. These two problems may be overcome by using a desirability coefficient.

A *desirability coefficient* (Harrington 1965) combines all response measurements into one measurement. The coefficient is scaled so that it has the value $D = 0$ if any undesirable condition occurs, and the value $D = 1$ when all the most desirable conditions occur. A function that has this property is the geometric mean for K responses.

$$D = (d_1 \times d_2 \times \cdots \times d_K)^{1/K},$$

where each d_i is a desirability measurement for the ith response. The desirability measurement for each response is also scaled from 0 to 1 by using a chart like Figure 2.6. The first axis is the desirability measurement from 0 to 1. The remaining axes are used for individual responses and may be scaled to reflect the importance of a response to the total desirability coefficient D.

For an experiment in which three responses—color (C), acidity (pH), and viscosity (V)—will be measured, we wish to optimize conditions for the preparation of a solution. The most desirable conditions are a blue solution

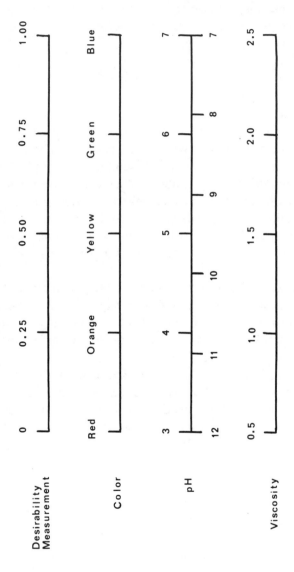

Figure 2.6.   Example of converting multiple responses to desirability measurements.

with a pH of 7.0 and a viscosity greater than 2.5 centipoise (cp). The color axis ranges from orange to blue and is equally spaced with increasingly desirable colors. The pH axis has two sides. The top scale ranges from 3.0 to 7.0, and the bottom scale ranges from 12.0 to 7.0. The basic side may be slightly preferred to the acidic side, so the scaling on the bottom is not scaled the same as the top. The viscosity axis ranges from 0.5 cp to 2.5 cp. If a viscosity measurement is $V < 0.5$ cp, the desirability coefficient will be assigned the value of $d_V = 0$, and any viscosity measurement of $V > 2.5$ cp will be assigned the value of $d_V = 1.0$. Suppose we apply a treatment that results in a green solution with a pH $= 9.0$ and a $V = 1.5$ cp. Then the desirability coefficient for the three responses is

$$d_C = 0.75, \quad d_{pH} = 0.60, \quad d_V = 0.50,$$
$$D = (0.75 \times 0.60 \times 0.50)^{1/3} = 0.61.$$

From this example we see several advantages of the desirability coefficient approach: (1) qualitative responses can be used; (2) responses with undesirable lower and upper limits can be used; (3) responses can have different scalings; (4) multiple responses can be transformed quickly to one measurement for optimization.

Other methods for multiple-response problems are available in Box et al. (1973) and Khuri and Conlon (1981). Both Box et al. and Khuri and Conlon consider the interdependencies among the multiple responses. Box et al. use an eigenvalue–eigenvector analysis to examine the effect of the relationship among observed measurements, whereas Khuri and Conlon minimize a weighted distance between the response function and the observed measurements.

2.7 Recording Data

Frequently, a major component of experimental error is due to improper recording of data. This component results from misreading responses, interchanging digits, using incorrect dilution factors, using faulty arithmetic, incorrectly transferring data, using incorrect units or scaling factors, omitting needed information, and writing illegibly. Data should be stored on computers, logbooks, and so on, in its most primitive form (eg, record counts/sec for both calibration and sample run). To minimize data recording errors, one should manipulate the basic data as little as possible by hand. This means that if a computer is available, it should be used to do all the arithmetic, scaling, and so on, and only few individuals should be involved in transferring basic experimental data to the final data file. Before the experiment begins, the following steps should be taken:

1. Design a data form.
2. Complete as much of the data form as possible. Record the explanatory variables, random order, and so on.
3. Have the person who will analyze the data review the data form.

After completing the experiment, you should recheck data for accuracy, misprints, and legibility. The data can then be examined by plotting the observations against each factor to detect any unusual data points, which can be reexamined for accuracy. Data should not be discarded just because it appears to be too high or too low. It may be necessary to quiz the chemist or technician who performed a particular experiment if there are any unusual conditions. Quizzing the chemist is a long shot because it is difficult to remember details of each experimental run performed some time ago. Therefore, it would be better to provide a space on the data form for comments about experimental conditions. The important axiom for data recording is to check, recheck, and check again.

3

Experimental Design and Analysis

Choosing an experimental plan depends on four main conditions: (1) number of factors and their levels, (2) number of blocking variables, if any, (3) number of replicate experimental units, and (4) methods of randomizing experimental units. To aid in choosing an experimental plan, nine plans commonly used by chemists will be described in the first nine sections of the chapter.

Completely randomized designs
Randomized block designs
Incomplete block designs
Factorial designs
Fractional factorial designs
Split-plot designs
Response surface designs
Mixture experimental designs
Computer-aided designs

Factorial, fractional factorial, and response surface designs will be discussed further in Chapters 4, 5, and 6.

The last five sections describe essential aspects of statistical analysis, accuracy, precision, and assumptions used to analyze and interpret experimental results. Many books have been written on statistical analysis, and it is our purpose not to duplicate these books but merely to provide the essential structure of analysis.

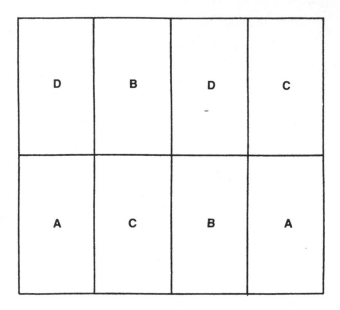

Figure 3.1.   Completely randomized design for a treatment factor with four levels A, B, C, and D. Each level is replicated twice and randomized over eight experimental units.

3.1 Completely Randomized Designs

Description

This design is for replicating a set of treatments on experimental units. Every experimental unit has the same probability of receiving any one of the treatments. This allocation is accomplished by assigning treatments to experimental units entirely at random. This design is most effective when all available experimental units are essentially homogeneous. An example of a completely randomized design is given in Figure 3.1.

Advantages

The number of treatments and replicates is limited only by the number of experimental units available. The number of experimental units for estimating the experimental error is maximum. The statistical analysis is simple even when the results from some of the experimental units are missing or rejected.

Disadvantages

The experimental error includes all variations among experimental units except those due to treatments. If substantial heterogeneity does occur among the experimental units, the completely randomized design may be inefficient.

Randomization

The t treatments that are to be replicated r times are allocated to rt experimental units entirely at random. To do the randomization, assign numbers $1, \ldots, rt$ to the experimental units, select rt random numbers, and rank these numbers from 1 to rt. Assign treatment 1 to experimental units corresponding to the ranks of the first r random numbers; assign treatment 2 to experimental units corresponding to the ranks of the second r random numbers, and so on.

References for Statistical Analysis

Snedecor and Cochran (1967), Steele and Torrie (1960), and Davies (1956).

3.2 Randomized Block Designs

Description

This design is for a set of t treatments to be replicated r times in a block of homogeneous experimental units. A blocking variable with b levels is used to block or group each rt experimental units, such that the experimental units in each block are as homogeneous as possible. The number of experimental units required is btr; for $r > 1$, this design is called a *completely randomized block design*. The object of this experimental plan is to keep the experimental error within each block as small as possible. Figure 3.2 shows a completely randomized block design for t $=$ 2, b $=$ 3, and r $=$ 2.

Advantages

More accurate results are usually obtained by blocking than with completely randomized designs. Any number of treatments and blocks may be used. The statistical analysis is straightforward. Mishaps that cause omission of a block or a treatment do not introduce complications. Variability among blocks is arithmetically removed from both the treatment effects and the experimental error.

Disadvantages

If variation among experimental units within a block is large, a large

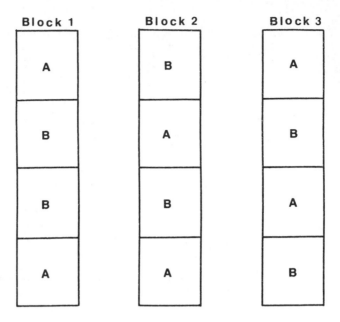

Figure 3.2.   Randomized block design for a treatment factor with two levels A and B. Each level is replicated twice and randomized over the four experimental units within each block.

error term results, and other designs to control a greater proportion of the variation should be used. For $r = 1$, any block by treatment interaction cannot be estimated and is included in the experimental error.

Randomization

When rt experimental units are assigned to each block, the replicate treatments are randomized in each block in the same manner as the completely randomized design. A new randomization is made for each block and, if possible, the blocks should be run in random order.

References for Statistical Analysis

Snedecor and Cochran (1967), Steele and Torrie (1960), and Davies (1956).

3.3 Incomplete Block Designs

Description

Sometimes experimental conditions will not permit a sufficient number of experimental units in each block to include every treatment. For example, t treatments are replicated r times, but no treatment appears more than once in each block, and some treatments are missing in each block. The blocking variable has b levels with k experimental units in each block so that $t > k$. Each pair of treatments appears together λ times. The five integer parameters t, r, b, k, and λ are not independent and are subjected to the following restrictions:

1. $N = tr = bk =$ total number of experimental units.
2. $\lambda (t - 1) = r (k - 1)$.
3. $t \leq b$.

Incomplete block designs do not exist for all parameter combinations. Existing designs are difficult to derive but are conveniently found in published tables. Tables of incomplete block designs are given in Cochran and Cox (1957), Fisher and Yates (1953), and Beyer (1966). The incomplete block design is illustrated in Figure 3.3.

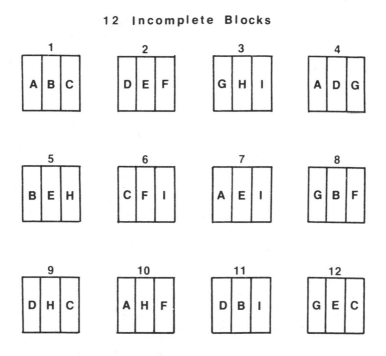

Figure 3.3. Incomplete block design for a treatment factor with nine levels ($t = 9$), A through I in 12 blocks ($b = 12$). Each block has three experimental units ($k = 3$) and each level is replicated four times ($r = 4$). Each pair of treatment levels appears together once ($\lambda = 1$).

Advantages

Incomplete block designs provide valid and efficient comparisons between treatments when all treatments cannot be tested under uniform conditions.

Disadvantages

Incomplete block designs do not exist for all cases we may wish to use. These designs are difficult to conduct and analyze. Adjusting for missing observations and other experimental complications is difficult.

Randomization

Select a tabulated incomplete block design that meets the experimental conditions. Assign the block levels randomly to the selected design and randomize the order of the blocks. Then assign the treatments randomly to the selected design's treatments.

Reference for Statistical Analysis

Cochran and Cox (1957), Davies (1956), Kempthorne (1952).

3.4   Factorial Designs

Description

In a factorial experiment, effects of a number of different factors are investigated simultaneously. The treatments or treatment combinations consist of all possible combinations of the levels from the different factors. For example, factors A and B have levels a_1, a_2, and b_1, b_2, b_3, respectively. The treatment combinations are (a_1, b_1), (a_1, b_2), (a_1, b_3), (a_2, b_1), (a_2, b_2), and (a_2, b_3). The number of experimental units required is the product of the number of levels for each factor and the number of times each treatment combination will be replicated. Figure 3.4 shows factorial designs for both qualitative and quantitative factors.

Advantages

Factorial designs are highly efficient because every observation provides information about all the factors in the experiment. When interactions (eg, when the effect of factor A depends on the levels of factor B) among the factors exist, a factorial design is necessary to avoid misleading conclusions. When no interactions exist, a factorial design gives the maximum efficiency in the estimation of the main effects of the factors. The conclusions hold over a wide range of conditions.

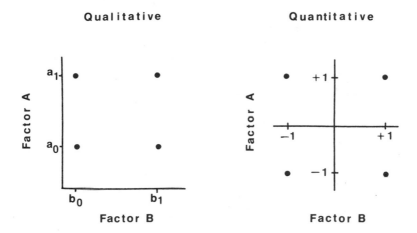

Figure 3.4.   Factorial designs for two factors A and B with each factor having two levels.

Disadvantages

When there are many factors, or each factor has many levels, a large number of experimental units is required. Also, when experimental units are heterogeneous, fractional factorial designs run in blocks may be more efficient.

Randomization

Randomize the total number of treatment combinations over the experimental units as in a completely randomized design.

References for Statistical Analysis

Snedecor and Cochran (1967), Steel and Torrie (1960), Davies (1956).

3.5   Fractional Factorial Designs

Description

Fractional factorial designs play a dual role in experimental design. First, fractions of the total number of treatment combinations of a factorial design are run in blocks when uniform conditions cannot be maintained during the factorial experiment. In this role, fractional factorial designs are a special case of incomplete block designs. Second, each fraction of a factorial design may stand as a complete experimental plan. In this role, information on certain effects (usually higher-order interactions) is sacrificed to gain

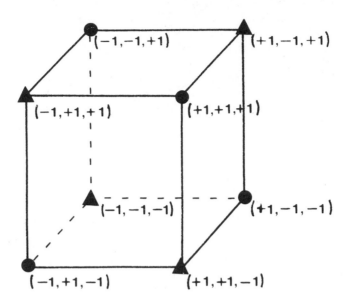

Figure 3.5. Two fractional factorial designs of a 2^3 factorial design. One fraction is represented by circles and the other fraction by triangles.

information on the effects of a large number of factors. Figure 3.5 shows how a 2^3 factorial design can be partitioned into two fractional factorial designs.

Advantages

Fractional factorial designs are very efficient for screening the effects of many factors by using a small number of experimental units.

Disadvantages

If interactions among the factors exist, it may not be possible to separate out their effects. Missing observations and other complications may make the statistical analysis impossible.

Randomization

List the total number of treatment combinations and then randomize them over the experimental units as in a completely randomized design. If more than one block is used, randomize within each block and run the blocks in random order.

References for Statistical Analysis

Davies (1956) and Kempthorne (1952).

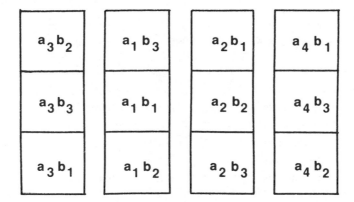

Figure 3.6.   Split-plot design for two factors A and B. The four levels of factor A are first randomized over the four sets of experimental units, then the three levels of factor B are randomized over the three experimental units within each set.

3.6   Split-Plot Designs

Description

Frequently, levels of one factor cannot be changed as rapidly (eg, waterbath temperatures) as those of other factors (eg, solution concentrations). In a split-plot design the levels of one or more factors, whole-factors, are kept at a fixed value for a set of experimental units while the levels of the remaining factors, subfactors, are randomized over the set of experimental units. Split-plot designs may be used when the levels of whole factors require larger amounts of experimental material for each experimental unit or when a more involved procedure is required to change whole-factor levels than to change subfactor levels. A split-plot design is shown in Figure 3.6 for a four-level whole-factor and a three-level subfactor.

Advantages

The experimental error is smaller for subfactors than for whole-factors. Therefore, a split-plot design can be used to study some factors with greater precision than others.

Disadvantages

Split-plot designs are often mistaken for factorial designs and incorrectly analyzed. In consequence, whole-factor errors are too small, and subfactor errors are too large. If, in a factorial experiment, all the main effects and interactions turn out to be highly significant, go back and examine the experimental design to see if it is really a split-plot design.

Randomization

Divide the experimental units into sets. The number of sets equals the number of whole-factor treatment combinations, and the number of experimental units within each set equals the number of subfactor treatment combinations. Randomization is done in two stages. First, randomize the treatment combinations of the whole-factors over the sets of experimental units. Second, randomize the treatment combinations of the subfactors within each set.

References for Statistical Analysis

Snedecor and Cochran (1967), Steel and Torrie (1960), Kempthorne (1952).

3.7   Response Surface Designs

Description

These designs are used to characterize a response that is approximated by a polynomial (usually low order) of quantitative factors. Chemists are often faced with (1) finding a region in the factor space where the response is highest (or lowest); and, (2) having found it, designing an experiment to map the response over that region. The first problem is examined in Chapter 5, and the second is dealt with in Chapter 6. Response surface designs are used to answer the second question and are usually designs that are factorials, fractional factorials, or fractional factorials with additional experimental units at the center and on the axes of the design (composite designs). Three criteria for design selection are the variance of estimated responses and coefficients, the bias of the approximating model, and the ability to block experimental units. Figure 3.7 is an illustration of a central composite design to study the response surface of two quantitative factors.

Advantages

Response surface designs are highly efficient designs for estimating approximating polynomial models. A variety of designs exists to satisfy different design criteria that an experimenter may need.

Disadvantages

Response surface designs and analyses are not used primarily for the

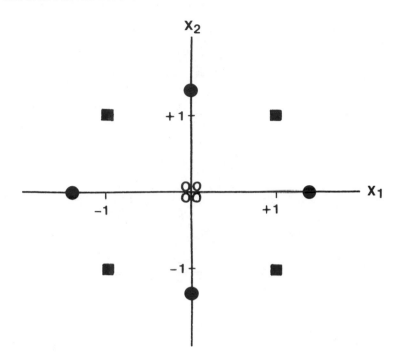

Figure 3.7. Central composite design to study the response surface for two quantitative factors. The solid squares are the factorial points, the solid circles are the axial points, and the open circles are the center points.

purpose of understanding the mechanism of the underlying system or process. However, these analyses may aid in determining that underlying mechanism. The design selection depends on the experimenter's ability to postulate an appropriate approximating model.

Randomization

The treatment combinations for the factors are randomized over the experimental units as in completely randomized designs. For designs that involve blocking, the randomization procedure for randomized block designs is followed. Methods for blocking central composite designs are given in Chapter 6.

References for Statistical Analysis

Myers (1971) and Davies (1956).

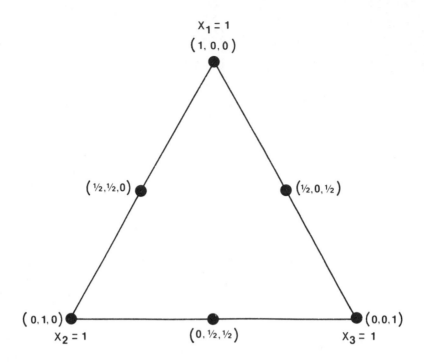

Figure 3.8.   Mixture design for three factors. All experimental points must lie on or inside the triangle and must satisfy the relationship $X_1 + X_2 + X_3 = 1$.

3.8   Mixture Experimental Designs

Description

Mixture designs are for experiments where the response depends only on the proportions of ingredients in a mixture rather than on their amount. For example, the two mixtures of iron and sulfur of (1 g, 1 g) and (10 g, 10 g) are both mixtures with 50% iron and 50% sulfur. The levels of each factor are the proportions of an ingredient, so each level is greater than or equal to zero, and the sum of the levels for each treatment combination is one. Because of these constraints, the factor space for K factors is defined by a regular $(K-1)$-dimensional simplex. The two-dimensional simplex in Figure 3.8 illustrates a mixture design for three factors.

Advantages

A variety of designs exists for modeling response surface functions over the simplex factor space.

Disadvantages
The design selection depends on the suitability of the postulated model for the response surface function.

Randomization
The treatment combinations are randomized over the experimental units as in completely randomized designs.

Reference for Statistical Analysis
Cornell (1981).

3.9   Computer-Aided Designs

Description
The majority of designs chemists will encounter are those discussed in the first eight sections. Occasionally, a situation arises in which special conditions are placed on the treatment combinations, or the experimental units are so expensive that the requirements of the standard designs cannot be met. For these situations a computer search for possible designs to satisfy certain design criteria can be made (Welch 1982; Cook and Nachtshein 1980; Mitchell 1974). The computer search finds treatment combinations that minimize variance criteria on either the estimated expected response or on coefficients of the response function. Figure 3.9 illustrates a computer-aided design that selected the best six points from a nine-point design.

Advantages
Computer-aided designs may provide the only possible method to design an experiment with restrictive conditions.

Disadvantages
When the number of experimental units is large, computer programs used to search for an appropriate design may require a mainframe computer to meet the storage and computational requirements. Also, because these designs often lack the symmetry of previously discussed designs, analysis of the response data will be more complex and will generally require the use of a computer.

Randomization
The treatment combinations are randomized over the experimental units as in completely randomized designs.

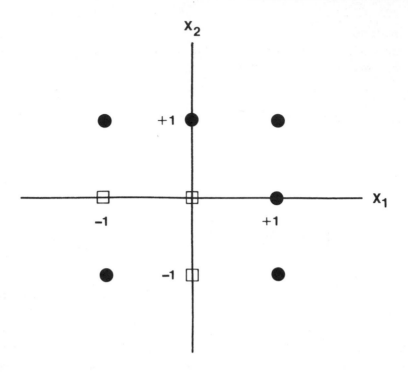

Figure 3.9. Computer-aided design that selected the best six points (solid circles) from a full three-level factorial design (solid circles and open squares) for two factors. The design points were selected based on a variance criterion called D-optimality.

Reference for Statistical Analysis

The statistical analysis of a computer-aided design is done by the methods of regression analysis (Draper and Smith 1981).

3.10   An Overview of Analyzing Experimental Results

The response data of an experiment consist of observed values of random variables. For example, response data on treatments A and B that are replicated r times in completely randomized experiment are observations of two random variables, each measured r times. The responses on experimental units for different treatment combinations represent realizations of different random variables. The realizations of each random variable are observed when replicate experimental units are measured.

Because the statement of objectives was formulated in terms of probability properties of random variables, it is now possible to apply probability theory to response data to investigate important questions related to these objectives. Are all probability distributions of the different random variables the same type: normal, Weibull, Poisson, binomial, and so on? Are all the random variables independent? Can parameters of the probability distributions be estimated, and if so, what are the variances of the estimators? Are the parameters that define the probability distributions, such as the means and variances, equal? The analysis of experimental results is the application of probability theory to the distributions of these random variables.

Statistical analysis of experimental responses involves choosing appropriate probability distribution models for random variables. These probability models are based on theoretical chemical and physical properties of the random variables and related previous experiments. For example, in kinetic studies we may be able to postulate that an expected response is due to either a first- or second-order reaction model based on stoichiometric equations.

Theoretical considerations also determine the type and range that response values can realize. Responses of random variables may be continuous or discrete; and their range of values may be all real numbers, only positive real numbers, a real number interval, positive and negative integers, a set of integers, rational numbers, and so on. These considerations are used to postulate the form of probability distribution models. The assumptions of postulated models are supported by the experimental plan and usually are verified to be reasonable by examining the experimental responses. For example, randomization is used in an experimental plan to support the assumption that responses are measured independently. Some response measurements cannot be randomized (eg, measurements in time) and are not independent. For these cases the correlation structure is part of the model assumptions, and these assumptions should be examined by using the experimental results.

The statement of objectives is frequently formulated in terms of the unknown parameters of the postulated probability model. Therefore, the objective of many experiments is to estimate these unknown parameters from response data and to find the variances of these estimators. The function of the response random variables used to estimate a parameter is called an *estimator*, and the numerical value of the estimator calculated from response data is called an *estimate*. Estimators of unknown parameters are frequently designated by placing a hat "$\hat{\ }$" over the parameter, and this is read "estimator of." For example, if the average of the responses (eg, $\bar{Y}$) is an estimator of the expected value of a random variable (eg, μ), the designation would be $\hat{\mu} = \bar{Y}$, and it would be read as $\bar{Y}$ is the estimator of

μ. Because estimators are functions of random variables, they are also random variables with probability distributions. An important property of an estimator is that its expected value should be equal to the parameter being estimated. If the estimator's expected value equals the parameter, the estimator is said to be *unbiased*; if they are not equal, the estimator is said to be *biased*. Therefore, the expected value of an estimator is a measure of the *accuracy* of the estimator. The *precision* of an estimator is measured by its standard deviation. Statements about estimators such as "an estimator has a high degree of precision," or "an estimator is very precise" mean that the standard deviation of the estimator is small. The accuracy and precision of some common estimators will be examined in the next section.

In addition to estimating parameters, comparisons of parameter values to each other or to a constant value are usually part of the experimental objectives. Do analytical methods A and B give the same expected response for analyzing manganese in steel? Here the hypothesis (null hypothesis) that the expected responses for the two analytical methods are equal would be tested against the alternative hypothesis that they are not equal. We use the *null hypothesis* in the context of a hypothesis under test of "no differences," and the *alternative hypothesis* as a hypotheses that "differences exist." Tests of hypotheses about parameters of probability distributions of the experimental responses are made by using test statistics.

A *test statistic* is a function of the response random variables that also has a probability distribution. The probability distribution will be one type if the null hypothesis is true, and another type if the alternative hypothesis is true. If the null hypothesis is assumed true and the evaluated test statistic is greater or less than expected (eg, greater than the 97.5% percentile value or less than the 2.5% value), then the null hypothesis is rejected; otherwise, the null hypothesis is not rejected. Note that the null hypothesis is never accepted, because a later experiment with more experimental units could have a higher probability of detecting differences and could lead to rejection of the null hypothesis. This principle of formulating a test statistic and then comparing it to percentile points of its probability distribution, assuming the null hypothesis is true, is the basis of the most common methods of statistical analysis: *t*-test, analysis of variance, and multiple comparisons.

3.11   Accuracy of an Estimator

Suppose the responses of the random variables $Y_1, Y_2, \cdots, Y_N$ are measured in an experiment. Assume that the random variables are indepen-

TABLE 3.1. Estimators of the population parameters from an i.i.d. sample of N responses

Population parameter	Estimator[a]	Expected value[b]	Variance of estimator[c]
μ	$\bar{Y}$	μ	σ^2/N
σ^2	S^2	σ^2	$2\sigma^4/(N-1)$
σ	S	$K\sigma$	$(1-K^2)\sigma^2$

[a] $\bar{Y} = \sum\limits_{i=1}^{N} Y_i/N$ and $S^2 = \sum\limits_{i=1}^{N} (Y_i - \bar{Y})^2/(N-1)$.

[b] $K = \sqrt{2}\ \Gamma(N/2)/\sqrt{N-1}\ \Gamma[(N-1)/2]$, where Γ is the gamma function (Abramowitz and Stegun 1972).

[c] The additional assumption that the responses are normally distributed is used to calculate the variances for S^2 and S.

dent, all with the same expected value, $E(Y_i) = \mu$, and variance, $\text{Var}(Y_i) = \sigma^2$. Three estimators of the parameters μ, σ^2, and σ are given in Table 3.1.

The notations $\bar{Y}$, S^2, and S rather than the hat notation are commonly used to denote estimators of the expected value, variance, and standard deviation, respectively, of a random variable. By examining the expected value of these estimators, we see that $\bar{Y}$ and S^2 are accurate or unbiased estimators for parameters μ and σ^2, but S is a biased estimator for σ. The biasness of the estimator S depends on the number of experimental units used to calculate the multiplier K. Table 3.2 shows the value of the K multiplier for $N = 2$ to 10. This table shows that biasness decreases rapidly for initial increases in N and then levels off.

The most common estimators in statistics are linear estimators of the form

$$\sum_{i=1}^{N} a_i Y_i$$

where each a_i is a constant value, and each Y_i is a random variable. The estimator $\bar{Y}$ is an example of a linear estimator where each constant has the same value $a_i = 1/N$. The expected value of a linear estimator can be calculated by the general formula

$$E\left(\sum_{i=1}^{N} a_i Y_i\right) = \sum_{i=1}^{N} a_i E(Y_i).$$

Table 3.2. Values of the multiplier K for the expected value and standard deviation of the estimator, S, calculated from a sample of N responses i.i.d. $N(\mu, \sigma^2)$

Sample size	K value	$1-K^2$
2	0.80	0.60
3	0.89	0.46
4	0.92	0.39
5	0.94	0.34
6	0.95	0.31
7	0.96	0.28
8	0.97	0.26
9	0.97	0.25
10	0.97	0.23

Another example of a linear estimator is the least squares estimator of the slope of a line. The assumed model for the expected value is $E(Y_i) = \bar{Y} + B(X_i - \bar{X})$ for the measured responses, Y_i, at the fixed values of the X_i's. The linear estimator of the slope B is

$$b = \sum_{i=1}^{N} a_i Y_i,$$

where

$$a_i = (X_i - \bar{X}) \Big/ \sum_{i=1}^{N} (X_i - \bar{X})^2.$$

We note that $\sum_{i=1}^{N} a_i = 0$ and $\sum_{i=1}^{N} a_i (X_i - \bar{X}) = 1$.

$$E(b) = \sum_{i=1}^{N} a_i E(Y_i) = \sum_{i=1}^{N} a_i [\bar{Y} - B(X_i - \bar{X})]$$

$$= \bar{Y} \sum_{i=1}^{N} a_i + B \sum_{i=1}^{N} a_i (X_i - \bar{X}) = B.$$

Therefore, the least squares estimator of the slope is an unbiased estimator.

The estimator, S^2, for σ^2 is not a linear estimator and requires more algebra to show that it is unbiased.

3.12   Precision of an Estimator

The precision of an estimator is its standard deviation. For example, the precision of $\bar{Y}$ is $\sigma / \sqrt{N}$. The term "standard error" is also frequently used for the standard deviation of $\bar{Y}$. Judging from the chemical literature, we believe that the standard deviation for a single observed value of a random variable Y is frequently confused with the standard deviation for the estimator $\bar{Y}$. A common practice is to report $\bar{Y} \pm S$ rather than $\bar{Y} \pm S/\sqrt{N}$. The estimator is usually the important value being reported, and the standard deviation of the estimator is its relevant precision. Standard deviations and variances of estimators are expressed in terms of the unknown population parameters and, therefore, must also be estimated. Keeping track of what is a parameter and what is an estimator of a parameter is important. For example, $\bar{Y}$ is an estimator of the parameter μ, and the variance of the estimator is $\text{Var}(\bar{Y}) = \sigma^2/N$. This variance, which is expressed in terms of the parameter σ^2, is estimated by S^2/N.

Before discussing methods for estimating variances of estimators, we will investigate nonindependent random variables. Let $f(y, z)$ be the joint probability density function of the random variables Y and Z, with $g(y)$ and $h(z)$ the respective marginal probability density functions (Hogg and Craig 1970). The random variables Y and Z are independent if and only if $f(y,z) = g(y)h(z)$. In many cases this equality does not hold, and the random variables are said to be *dependent*. A measure of this dependency is *covariance*, defined by

$$\text{Cov}(Y, Z) = E[(Y - \mu_Y)(Z - \mu_Z)]$$

$$= E[YZ] - \mu_Y \mu_Z.$$

The value of $\text{Cov}(Y,Z)$ can be either positive or negative. If the random variables Y and Z are independent, then $E(YZ) = E(Y)E(Z)$, and the covariance is zero. However, the reverse is not true; that is, if $\text{Cov}(Y,Z) = 0$, we cannot infer that Y and Z are independent random variables.

A related statistic is the correlation coefficient defined by

$$\rho = \frac{\text{Cov}(Y, Z)}{[\text{Var}(Y)\ \text{Var}(Z)]^{1/2}}.$$

The correlation coefficient ranges between -1 and $+1$ and is a measure of the linear relationship between two random variables. This is a mathematical relationship and does not necessarily imply any cause-and-effect relationship. For example, over a period of years the correlation coefficient between ministers' salaries and the consumption of liquor was highly positive. However, this result does not mean that ministers increased their consumption of liquor. During this same period there was a general upward trend in wages; thus, the increased purchases of liquor could be related to the increase in purchasing power of the general population. Successful interpretation of the correlation coefficient requires familiarity with the field of application as well as its mathematical properties.

To estimate the covariance and the correlation coefficient, we define the notation for corrected cross products, S_{YZ} , for N paired random variables (Y_1,Z_1), (Y_2,Z_2), $\cdots$, (Y_N ,Z_N) by

$$S_{YZ} = \sum_{i=1}^{N} (Y_i - \bar{Y}) (Z_i - \bar{Z}).$$

Similarly, this notation is also used for the corrected sum of squares of a single random variable as

$$S_{YY} = \sum_{i=1}^{N} (Y_i - \bar{Y})^2.$$

Using this notation, we define estimators for $\mathrm{Cov}(Y,Z)$ and ρ by

$$\widehat{\mathrm{Cov}}(Y, Z) = \frac{S_{YZ}}{N - 1} \quad \text{and} \quad \hat{\rho} = \frac{S_{YZ}}{(S_{YY}S_{ZZ})^{1/2}} ,$$

where the hat ``$\hat{}$'' is read ``estimator of.''

The variance of a linear combination of random variables can now be formulated in terms of individual variances and covariances between random variables. Let $Y_1,Y_2, \cdots , Y_N$ denote N random variables, with a_1, $a_2, \cdots , a_N$ denoting real constants in the linear combination

$$Z = \sum_{i=1}^{N} a_i Y_i.$$

Then the variance of Z is

$$\mathrm{Var}(Z) = \sum_{i=1}^{N} a_i^2 \mathrm{Var}(Y_i) + 2\sum\sum_{i < j} a_i a_j \, \mathrm{Cov}(Y_i, Y_j).$$

This is an extremely useful formula, which is illustrated with the following examples:

Example 3.1

Let Y_i , $i = 1, \cdots , N$, be N independent random variables (ie, $\mathrm{Cov}(Y_i , Y_j)=0$ for every $i \neq j$ pair). Find the variance of the mean estimator, $\bar{Y}$, assuming a constant variance; that is, $\mathrm{Var}(Y) = \mathrm{Var}(Y_1)= \cdots = \mathrm{Var}(Y_N)$. The constants $a_i = 1/N$ for all values of $i = 1, \cdots , N$.

$$\mathrm{Var}(\bar{Y}) = \sum_{i=1}^{N} \left(\frac{1}{N}\right)^2 \mathrm{Var}(Y_i)$$

$$= \sum_{i=1}^{N} \frac{\mathrm{Var}(Y)}{N^2} = \frac{N \, \mathrm{Var}(Y)}{N^2} = \frac{\mathrm{Var}(Y)}{N}.$$

This variance would be estimated by substituting S^2 for $Var(Y)$.

Example 3.2

Let Y_i, $i = 1, \cdots, N$, be N independent random variables measured at the fixed values, X_i (again $Cov(Y_i, Y_j) = 0$ for every $i \neq j$ pair). Find the variance of the slope estimator, $\hat{B} = b$, for the line $E(Y_i) = \bar{Y} + B(X_i - \bar{X})$. Assume $Var(Y) = Var(Y_1) = \cdots = Var(Y_N)$ and recall that $a_i = (X_i - \bar{X})/S_{XX}$. Then by the variance formula for a linear combination,

$$Var(b) = \sum_{i=1}^{N} \frac{Var(Y_i)(X_i - \bar{X})^2}{S_{XX}^2}$$

$$= \frac{Var(Y) S_{XX}}{S_{XX}^2} = \frac{Var(Y)}{S_{XX}}.$$

The variance, $Var(Y)$, would be estimated by dividing the sum of squares of the residuals,

$$SSR = \sum_{i=1}^{N} [Y_i - \bar{Y} - b(X_i - \bar{X})]^2,$$

by its corresponding degrees of freedom [ie, $\hat{Var}(Y) = SSR/(N-2)$].

Example 3.3

	TPM	Nicotine	Water	Tar
	44.43	2.66	5.78	35.99
	42.37	2.52	5.46	34.39
	42.83	2.60	5.36	34.87
	43.16	2.62	5.55	34.99
	44.91	2.58	6.06	36.27
	44.91	2.63	5.65	36.63
	44.12	2.64	5.24	36.24
	43.82	2.57	5.53	35.72
$\bar{Y} =$	43.82	2.60	5.58	35.63
$S^2 =$	0.908	0.002	0.065	0.634

Nicotine, water, and total particulate matter are measured on eight cigarette samples. The amount of tar is determined by subtracting the amount of nicotine and water from the total particulate matter. What is the variance of the tar measurement (note that we want the variance of one measurement not the variance of the average value)? Let T, TPM, N, and W be random variables representing tar, total particulate matter, nicotine, and water, respectively. The values from eight cigarettes are given in milligrams delivered per cigarette.

The tar random variable is the linear combination $T = TPM - N - W$, and its variance is calculated by

$$Var(T) = Var(TPM) + Var(N) + Var(W) - 2Cov(TPM,N)$$
$$- 2Cov(TPM,W) + 2Cov(N,W).$$

From the data, the covariance estimates are $\hat{Cov}(TPM,N) = 0.023$, $\hat{Cov}(TPM,W) = 0.147$, and $\hat{Cov}(N,W) = 0.0002$. Substituting both the variance and covariance estimates in the variance formula, we calculate an estimate of Var(T) by

$$\hat{Var}(T) = 0.908 + 0.002 + 0.065 - 0.046 - 0.294 + 0.0004$$

$$= 0.635.$$

It is interesting to compare the estimate of Var(T), calculated by using the variance formula for linear estimators, with the estimate made from the data (ie, $S^2 = 0.634$). The two estimates are mathematically equivalent but differ slightly due to roundoff error.

Frequently, chemists need variances of functions of random variables other than linear functions. The reason is that the final concentration of an unknown may be determined by a series of arithmetic steps: subtracting tare weights, multiplying by the sample volume, dividing by a standard, and so on. Let Z be a continuous random variable related to several random variables $Y_1, Y_2, \cdots, Y_N$, by the function

$$Z = F(Y_1, Y_2, \cdots, Y_N).$$

The variance of Z can be approximated by expanding F in a Taylor's series about the expected values of the Y's and neglecting all terms except the first-order-derivative terms. Approximate the mean of Z by $F(\mu_1, \cdots, \mu_N)$, subtract this value from both sides, square the difference, and take expected values. The resulting approximation is

$$Var(Z) = \sum_{i=1}^{N} \left(\frac{\partial F}{\partial Y_i}\right)^2 Var(Y_i) + 2\sum_{i}\sum_{<j} \left(\frac{\partial F}{\partial Y_i}\right)\left(\frac{\partial F}{\partial Y_j}\right) Cov(Y_i, Y_j).$$

The derivatives are evaluated by first replacing the random variables with their expected values (eg, $E(Y)$'s). These expected values are then estimated by using unbiased estimators, usually forms of $\bar{Y}$'s, S^2's, and estimated covariances.

The bias due to approximating the mean of Z by the function of the means can be quantified by examining the second partial derivatives. For a function of two variables Y and W, this bias is approximated by

$$E[F(Y,W) - F(\mu_Y, \mu_W)]$$

$$= \frac{1}{2}\left[\frac{\partial^2 F}{\partial Y^2} \text{Var}(Y) + 2 \frac{\partial^2 F}{\partial Y \partial W} \text{Cov}(Y,W) + \frac{\partial^2 F}{\partial W^2} \text{Var}(W) \right].$$

Example 3.4

The estimated coefficient of variation,

$$\hat{CV} = \frac{S}{\bar{Y}} = F(S,\bar{Y}),$$

is frequently used as a measure of error. What is the variance of $\hat{CV}$ if from five measurements the mean and variance estimates are $\bar{Y} = 1.0$ and $S^2 = 0.01$? We can show that $\text{Cov}(\bar{Y},S) = 0$ (Hogg and Craig 1970), and, by the variance approximation formula,

$$\text{Var}(\hat{CV}) = \left(\frac{\partial \hat{CV}}{\partial S}\right)^2 \text{Var}(S) + \left(\frac{\partial \hat{CV}}{\partial \bar{Y}}\right)^2 \text{Var}(\bar{Y}).$$

The derivatives are $\partial \hat{CV}/\partial S = 1/\bar{Y}$, and $\partial \hat{CV}/\partial \bar{Y} = -S/\bar{Y}^2$. Using Table 3.1, we evaluate these derivatives at $E(\bar{Y}) = \mu$ and $E(S) = K\sigma$, which are estimated by $\hat{E}(\bar{Y}) = \bar{Y}$ and $\hat{E}(S) = K(S/K) = S$ (eg, S/K is the unbiased estimator of σ).

$$\text{Var}(\hat{CV}) = \frac{\text{Var}(S)}{\bar{Y}^2} + \frac{S^2 \text{Var}(\bar{Y})}{\bar{Y}^4}.$$

$\text{Var}(S)$ and $\text{Var}(\bar{Y})$ can also be found in Table 3.1, which leads to

$$\hat{\text{Var}}(\hat{CV}) = \frac{(1 - K^2)S^2}{\bar{Y}^2} + \frac{S^4}{N\bar{Y}^4}.$$

Using the estimated mean and variance values, we obtain the estimated $\text{Var}(\hat{CV})$ value as

$$\text{Var}(\hat{CV}) = \frac{(1 - (0.94)^2)(0.01)}{(1.0)^2} + \frac{(0.01)^2}{(5)(1.0)^4}$$

$$= 0.001184,$$

$$\sqrt{\text{Var}(\hat{CV})} = 0.0344.$$

The expected value of $\hat{CV}$ is approximated by $F(\mu_S, \mu_{\bar{Y}}) = K\sigma/\mu$. The difference between the true expected value and this approximation represents the bias. This bias can be estimated by using the approximating bias formula with $\partial^2\hat{CV}/\partial S^2 = 0$, $\text{Cov}(\bar{Y},S) = 0$, and $\partial^2\hat{CV}/\partial\bar{Y}^2 = 2S/\bar{Y}^3$. Again, these derivatives would be evaluated at $E(S)$ and $E(\bar{Y})$, and then unbiased estimators of these expected values would be used.

$$\hat{\text{Bias}} = \frac{2S^3}{2N\bar{Y}^3} = \frac{(0.1)^3}{5(1.0)^3} = 0.0002.$$

Therefore, the estimated relative error ($\hat{CV} \times 100\%$) for the five measurements is 10% with a standard deviation of 3.4% and a bias of 0.02%.

Example 3.5

The normality of an acid solution can be found by measuring the volume required to neutralize one gram-milliequivalent (0.0400 g) of pure NaOH. Estimate the normality of the acid solution and the standard error of the estimate. The following data represents the milliliters required to neutralize five samples of pure NaOH.

	NaOH (gm)	Acid (ml)
	2.510	66.5
	2.495	62.0
	2.484	75.6
	2.499	64.1
	2.496	55.5
average =	2.497	64.74
variance =	8.67E−5	53.58
covariance =	−0.026	

The normality of the acid solution is determined by

$$N = \frac{\text{grams of NaOH}}{(\text{ml of acid})(0.0400)}.$$

Let W be the random variable representing the weight of NaOH, and let V

be the random variable representing the volume of acid. Normality will be estimated by

$$\hat{N} = \frac{\overline{W}}{\overline{V} \times (0.0400)} = \frac{2.497}{(64.74)(0.0400)} = 0.964.$$

The variance of the normality estimate is

$$\text{Var}(\hat{N}) = \frac{\text{Var}(\overline{W}/\overline{V})}{(0.0400)^2},$$

where the numerator is approximated by

$$\hat{\text{Var}}\left(\frac{\overline{W}}{\overline{V}}\right) = \left(\frac{\overline{W}}{\overline{V}}\right)^2 \left(\frac{S_{\overline{W}}^2}{\overline{W}^2} + \frac{S_{\overline{V}}^2}{\overline{V}^2} - \frac{2\hat{\text{Cov}}(\overline{W},\overline{V})}{\overline{W}\,\overline{V}}\right),$$

Substituting the average values, $\overline{W}$ and $\overline{V}$, and their variances and covariance (ie, divide the variances and covariance by the sample size) gives

$$\hat{\text{Var}}\left(\frac{\overline{W}}{\overline{V}}\right) = \left(\frac{2.497}{64.74}\right)^2 \left[\frac{1.734\text{E}-5}{(2.497)^2} + \frac{10.716}{(64.74)^2} - \frac{2(-0.005)}{(2.497)(64.74)}\right]$$

$$= 3.900\text{E}-6,$$

$$\hat{\text{Var}}(\hat{N}) = 2.438\text{E}-3.$$

Therefore, the normality is estimated as $\hat{N} = 0.964$, and one standard deviation of the normality is estimated as 0.049.

The formulas for variance approximations are given in Table 3.3 for some commonly encountered functions. An excellent discussion of the propagation of error formulas is given by Ku (1966).

TABLE 3.3. Variance approximation formulas for functions of random variables

Function	Variance approximation[a,b]
$aY + bZ$	$a^2 V(Y) + b^2 V(Z) + 2ab C(Y,Z)$
$Y * Z$	$E(Z)^2 V(Y) + E(Y)^2 V(Z) + 2E(Y)E(Z)C(Y,Z)$
Y/Z	$[E(Y)/E(Z)]^2 [V(Y)/E(Y)^2 + V(Z)/E(Z)^2 - 2C(Y,Z)/E(Y)E(Z)]$
$\sqrt{Y}$	$V(Y)/4E(Y)$
$\ln(Y)$	$V(Y)/E(Y)^2$
$\exp(Y)$	$V(Y)\exp[2E(Y)]$

[a] $E(Y)$ and $V(Y)$ are the mean and variance of a random variable, and $C(Y,Z)$ is the covariance between two random variables.
[b] Small letters a and b represent constants.

3.13   Confidence Intervals

Frequently, chemists wish to express how well they have estimated an unknown parameter value. One method is to express a confidence interval for the parameter value. A *confidence interval* represents an interval in which the true parameter value would occur with a fixed probability if the experiment were repeated many times under exactly the same conditions. For a parameter θ, two endpoints t_1 and t_2 need to be determined from the response random variables to calculate the probability

$$Pr(t_1 < \theta < t_2) = 1 - \alpha.$$

The interval between t_1 and t_2 is called a confidence interval. The fixed probability values of $1 - \alpha$ are known as *confidence levels* or *confidence coefficients*, and the most commonly used values are 0.90, 0.95, and 0.99. Confidence intervals for a parameter are usually constructed based on the distributional properties of its estimator, which may produce either symmetrical or nonsymmetrical intervals about its estimated value.

Confidence intervals do not make any statement about the interval in which individual responses will occur if the experiment is repeated. Statements about the proportion of responses occurring in an interval are called *tolerance intervals* (Weissberg and Beatty 1960).

To demonstrate the construction of a confidence interval, we first put an estimator in a standardized form by subtracting its expected value and dividing by its standard deviation:

$$t = \frac{\text{estimator} - E(\text{estimator})}{\sqrt{\hat{\text{V}}\text{ar}(\text{estimator})}}.$$

The value of this *t*-statistic will vary from random sample to random sample because both the estimate and its estimated variance will vary. For experiments with replicated measured random variables that are independent and have identically normal distributions, the distribution of the *t*-statistic is known to be Student's *t*-distribution. Student was the nom de plume of William Sealy Gosset (1876–1937), an English chemist and statistician who worked for the Guinness firm of brewers. The *t*-distribution has a shape similar to the normal distribution, with its width depending on a parameter called degrees of freedom (df). The degrees of freedom is related to the sample size and the number of parameters estimated to calculate the variance of the estimator.

$$\text{df} = \text{sample size} - \text{number of estimated parameters}.$$

Figure 3.10 illustrates the *t*-distribution for df = 2, 4, and 16.

For a fixed sample size, the *t*-statistic will have a probability of $1 - \alpha$ of occurring between the values $-t_{\alpha/2}$ and $+t_{\alpha/2}$; that is, $\Pr(-t_{\alpha/2} < t < +t_{\alpha/2}) = 1 - \alpha$. Substituting the expression for the *t*-statistic and rearranging terms, we obtain the confidence limits for the expected value of the estimator

$$\text{estimator} \pm t_{\alpha/2} \sqrt{\hat{\text{V}}\text{ar(estimator)}}.$$

An important application of confidence intervals is for estimators that are a linear sum of independent random variables identically distributed as normal random variables. In such cases the linear sum is also distributed as a normal random variable.

Example 3.6

Let Y_i, $i = 1, \cdots, N, \sim$ i.i.d. $N(\mu, \sigma^2)$. Then $\bar{Y}$ is a linear estimator for the population mean. The estimated variance of the estimator is S^2/N, which is calculated based on the estimated value of the population mean. Therefore, the confidence interval for μ with a confidence coefficient of $1 - \alpha$ is

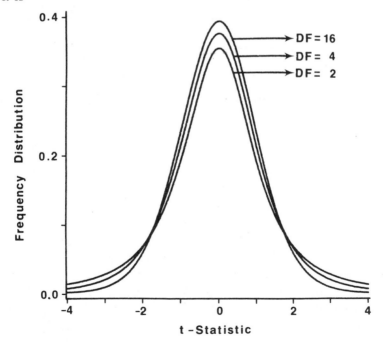

Figure 3.10. Probability frequency distribution for Student's t-Statistic for degrees of freedom DF = 2, 4, and 16.

$$\bar{Y} \pm t_{\alpha/2} \frac{S}{\sqrt{N}} \, .$$

The t-statistic value will have df $= N - 1$, and it can be computed by using the BASIC program in Table A.1 (see the appendix: BASIC Programs).

Example 3.7

Let $Y_i \sim$ i.i.d. $N(E(Y), \sigma^2)$, measured at fixed values of X_i, $i = 1,2, \cdots , N$. Let $E(Y)$ be modeled by the line $E(Y) = \bar{Y} + B(X_i - \bar{X})$. The linear estimator for the slope, b, has a normal distribution with an estimated variance, given in Example 3.2 in the previous section. The confidence interval for parameter B with $1 - \alpha$ confidence coefficient is

$$b \pm t_{\alpha/2} \sqrt{\frac{SSR}{(N-2) \, S_{xx}}} \, .$$

The appropriate degrees of freedom for the t-statistic are

$$\text{df} = N - 2,$$

because the variance for estimator b is based on two-parameter estimates (ie, intercept and slope).

Confidence intervals are related to testing the hypothesis that a parameter value is equal to a specified value, usually zero. The value of α in the confidence coefficient is the significance level, or the probability of rejecting the null hypothesis when it is true (type I error). If a confidence interval for a parameter includes a specified value, then the null hypothesis that the parameter value is equal to the specified value cannot be rejected. If the specified value falls outside the confidence interval, then the conclusion is that the null hypothesis is rejected. Suppose five identical solutions are analyzed for sodium cyanide by the Liebig method, with the following results: 0.122, 0.128, 0.119, 0.124, and 0.131 g NaCN. The 95% confidence interval for the expected concentration is 0.125 ± 0.006. Prior to the analysis, the null hypothesis that the expected concentration is not significantly different from 0.115 g was stated. The experimental results show that the true concentration is significantly different from 0.115 g at the 5% significance level because 0.115 is not in the 95% confidence interval. The statement ''significantly different at the 95% confidence interval'' is not correct and should not be used because it confuses hypothesis testing with parameter estimation.

3.14 Independence, Homogeneity, and Normality

Three assumptions used in many statistical analyses are the following:
1. Responses are independent.
2. Responses have equal variances.
3. Responses have a normal distribution.

The effect of invalid assumptions and the methods to test these assumptions will be examined.

Independent responses

The most common cause of nonindependent responses is that they are collected sequentially in time or in a systematic manner. These methods usually introduce positive correlation among the observations. Positive correlation among the observations causes too many significant results if either a t-test or an F-test in analysis of variance is used to test for treatment differences. Correlation often goes undetected because its presence is difficult to detect by inspection of the data. Occasionally, correlation cannot be avoided, as in the case when the responses are taken in a time series. Time series analysis, a special area of statistics, incorporates correlation structure into the model used to analyze the data (Box and Jenkins 1970). For the majority of experiments run in the chemistry laboratory, the most effective precaution against correlation is randomization.

Homogeneity of Variance

Frequently, new methods are compared against standard methods for analyzing chemical compounds with the hope of finding a faster or cheaper analysis. We should expect the precision of the standard method and the precision of the new method to differ. If equal replications are used for each method, no effects or moderate effects on the statistical comparison will occur. However, in general, nonhomogeneity can cause either too many or too few significant differences.

Nonhomogeneity can be classified as either *irregular* or *regular*. Irregular cases occur when a few treatments have variances much larger than the rest. These treatments can be omitted from the main analysis, and inferences about the set-aside treatments are made by inspection. These particular treatments may be considered as outliers. Suspected nonhomogeneity can be examined by calculating the estimated mean and variance for each treatment from replicated experimental units. Plotting variances versus means may detect outlier treatments or indicate the existence of a functional relationship (usually the variance increases as the mean increases).

The regular cases of nonhomogeneity are characterized by a functional relationship between variance and the expected value of the responses. For example, the variance equals the expected value for Poisson count data. We can transform responses on a new scale of measurement so that the variance is independent of the expected value. Additionally, this transformation will cause the transformed responses to be approximately normally distributed. Suppose that the variance of a response is related to the expected value, $\text{Var}(Y) = g[E(Y)]$, and $Z = h(Y)$ is the transformed value of a response. Then by a first-order Taylor series,

$$Z = h[E(Y)] + h'[E(Y)][Y - E(Y)],$$

$$\text{Var}(Z) = \{h'[E(Y)]\}^2 E[Y - E(Y)]^2.$$

By definition, $E[Y - E(Y)]^2 = \text{Var}(Y) = g[E(Y)]$; hence,

$$\text{Var}(Z) = \{h'[E(Y)]\}^2 g[E(Y)],$$

where "'" represents the derivative with respect to Y.

For Var (Z) to be independent of $E(Y)$, the function $h(Y)$ is chosen so that the right-hand side of the equation is a constant, which makes $h[E(Y)]$ the indefinite integral:

$$h[E(Y)] = \int \frac{d[E(Y)]}{\sqrt{g[E(Y)]}}.$$

For the Poisson distribution, $\text{Var}(Y) = E(Y)$, so that $h[E(Y)] = \sqrt{E(Y)}$, or $Z = \sqrt{Y}$. The variance on the transformed scale is

$$\text{Var}(Z) = \{h'[E(Y)]\}^2 g[E(Y)] = \left[\frac{1}{2\sqrt{E(Y)}}\right]^2 E(Y) = \frac{1}{4}.$$

Some useful transformations for data are given in Table 3.4.

TABLE 3.4. Transformations to correct for homogeneity and approximate normality

Data type	Transformation
Counts (Y)	$\sqrt{Y}$
Small counts (Y)	$\sqrt{Y + 1}$ or $\sqrt{Y} + \sqrt{Y + 1}$
Proportions ($0 < P < 1$)	arcsin ($\sqrt{P}$)
Variance = (mean)^2	ln(Y)
Correlation coefficient (r)	$0.5[\ln(1 + r) - \ln(1 - r)]$

An alternative method for both irregular and regular nonhomogeneous data is to use a weighting function. This method is used when the data is analyzed by the method of least squares. The weights are usually equal to the reciprocal of the variance estimate for each response (Draper and Smith 1981).

Normality

The assumption that responses are distributed as a normal distribution is frequently made to calculate confidence intervals, test for significant effects, or make additional data comparisons. The normality assumption is usually made because (1) extensive statistical methods and accurate tables have been derived for the normal distribution; (2) many normal results hold for nonnormal populations; (3) nonnormal responses may be transformed to induce normality; and (4) measurements that are considered the sum of many random variables (eg, error terms can be considered the sum of many small chance effects) can be considered approximately normal no matter what the underlying distribution if its variance is finite [see the Central Limit Theorem in Hogg and Craig (1970)].

Responses should be examined to verify that the normality assumption is reasonable. For example, the range of a normal response is $-\infty < Y < +\infty$, so responses such as proportions, $0 < P < 1$, or count data, $C = 0, 1, 2, \cdots$, must be transformed to be approximately normal, or a statistical analysis based on the appropriate probability distribution must be used. Statistical analysis developed on the exact probability distribution would be a more efficient analysis. However, in practice this approach is seldom attempted by chemists because the exact probability distribution is not known, or a more sophisticated analysis is considered too complicated.

Two statistical tests available for the null hypothesis "H_0: responses are normally distributed" versus the alternative hypothesis "H_1: responses are not normally distributed" are the Anderson–Darling test (Pettitt 1977) and a test based on the skewness and kurtosis statistics (Bowman and Shenton 1975). Other tests are available (ie, Shapiro–Wilk statistic), but these two tests have some computational advantages and similar probabilities of detecting nonnormality. A BASIC program for calculating the statistics for the two tests is given in Table A.2 (see the appendix: BASIC Programs).

The algorithm for calculating the Anderson–Darling statistic is the following:

1. Order the data: $Y_1 < Y_2 < \cdots < Y_N$. The sample size should be $N \geq 5$.
2. Standardize the ordered data by the estimated mean and standard deviation: $Z_i = (Y_i - \bar{Y})/S$.
3. Calculate the percentile points, P_i , using $N(0,1)$ probability distribution, by $P_i = \mathrm{Pr}(Z \leq Z_i)$.
4. Calculate the Anderson–Darling statistic:

$$A^2(N) = -\frac{1}{N}\left\{\sum_{i=1}^{N} (2i-1)\left[\ln(P_i) + \ln(1-P_{N+1-i})\right]\right\} - N.$$

5. Compare $A^2(N)$ with the percentile points for 0.05, 0.10, 0.15, 0.85, 0.90, 0.95, 0.975, and 0.99. The null hypothesis of normality is rejected at a selected significance level$-\alpha = 0.01, 0.025, 0.05, 0.10, 0.15-$if $A^2(N)$ is either less than the percentile points for α, or if $A^2(N)$ is greater than the percentile points for $1 - \alpha$.
6. The percentile points $a(N)$ are calculated by the formula

$$a(N) = A\left(1.0 + \frac{B}{N} + \frac{C}{N^2}\right).$$

where the coefficients A, B, and C are given in Table 3.5.

Another useful test whenever the question of normality arises is based on the sample skewness, $\sqrt{b_1}$, and kurtosis, b_2, statistics. The *skewness* of a distribution is the standardized third moment and has a value of 0 for

TABLE 3.5. Coefficients used to calculate percentile values $a(N) = A (1.0 + B/N + C/N^2)$ for the Anderson–Darling statistic (by kind permission A.N. Pettitt, 1977, *J. R. Statist Soc. C* 26:156–161)

Percentile points	A	B	C
0.05	0.1674	−0.512	2.10
0.10	0.1938	−0.552	1.25
0.15	0.2147	−0.608	1.07
0.85	0.5597	−0.749	−0.59
0.90	0.6305	−0.750	−0.80
0.95	0.7514	−0.795	−0.89
0.975	0.8728	−0.881	−0.94
0.99	1.0348	−1.013	−0.93

symmetrical distributions. Distributions with large right tails are likely to have large positive skewness values, and those with large left tails are likely to have large negative values. However, the skewness value may be zero for nonsymmetrical distributions, so care must be exercised in interpreting its value. The *kurtosis* of a distribution is the standardized fourth moment and is more difficult to interpret. Kurtosis has been used as a measure of peakedness, because the kurtosis of a normal distribution is 3, and distributions with kurtosis values less than 3 are flatter (eg, the kurtosis of the uniform distribution is 1.8), whereas those distributions with kurtosis values greater than 3 are pointier (eg, the kurtosis of the logistic distribution is 4.2). Because peakedness has a vague definition, kurtosis may be thought of as another quantity that characterizes a distribution without necessarily giving it a geometric interpretation.

Bowman and Shenton's test is useful for sample sizes $20 \leq N \leq 1000$, which covers many of the sample sizes encountered in practice. The algorithm for the Bowman–Shenton test is the following:

1. Calculate the sample moments M_r, for $r = 2,3,4$, defined by

$$M_r = \sum_{i=1}^{N} \frac{(Y_i - \bar{Y})^r}{N} .$$

2. Calculate the sample skewness and kurtosis:

$$\sqrt{b_1} = \frac{M_3}{M_2^{3/2}}, \quad b_2 = \frac{M_4}{M_2^2} .$$

3. Plot the values of $\sqrt{b_1}$ and b_2 on the appropriate contours in Figure

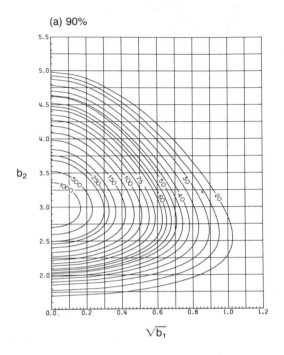

(a) 90%

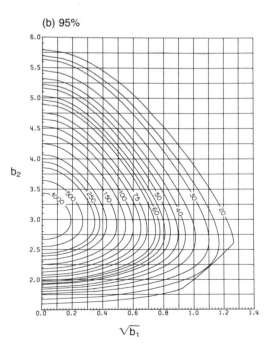

(b) 95%

(c) 99%

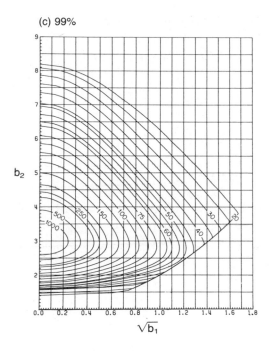

Figure 3.11.   Sample skewness and kurtosis contours for normality test; N = 20(5)65, 75, 85, 100, 120, 150, 200, 250, 300, 500, 1000. (With kind permission of the Biometrika Trustees, K. O. Bowman, and L. R. Shenton. 1975, *Biometrika* 61: 243–250).

3.11 for significance levels = 0.10, 0.05, and 0.01. If the value of $\sqrt{b_1}$ is negative, then use its absolute value (ie, the positive value).

4. If the value of $(\sqrt{b_1}, b_2)$ falls outside the contour for the given sample size, then reject the null hypothesis of normality.

Example 3.8

Using neutron activation, we analyzed a sample of 25 animal hairs for chromium. The following results were obtained as measured in ppm:

14.9	4.7	6.1	3.9	5.7
2.2	2.1	4.5	2.3	4.5
2.6	10.6	4.1	10.2	6.4
1.3	6.1	12.7	1.9	2.3
11.3	3.9	3.7	11.4	0.7

These data were tested for normality using the BASIC program listed in Table A.2. The resulting output is

N = 25   MEAN = 5.604   STANDARD DEVIATION = 3.958059
SKEWNESS = 0.8976286   KURTOSIS = 2.653929
ANDERSON–DARLING STATISTIC = 1.198319
PERCENTILE POINTS FOR ANDERSON–DARLING STATISTIC
LOWER-TAIL PERCENTILE POINTS
5%: 0.1645341   10%: 0.1899085   15%: 0.2098461
UPPER-TAIL PERCENTILE POINTS
85%:   0.542403   90%:   0.610778   95%:   0.7264355   97.5%:
0.8407298   99%: 0.991330

Comparing the Anderson–Darling statistic with the percentile points, we see that normality would be rejected at the 1% significance level. Plotting the skewness and kurtosis values on the Bowman–Shenton contours indicates that normality would not be rejected. For this set of data, the Anderson–Darling test is better able to detect non-normality than the Bowman–Shenton test.

If the data is transformed by taking the natural logarithm of the values and is resubmitted to the BASIC program, the following output is obtained:

N = 25   MEAN = 1.467769   STANDARD DEVIATION = 0.7662091
SKEWNESS = −0.3165532   KURTOSIS = 2.659723
ANDERSON–DARLING STATISTIC = 0.3054447
PERCENTILE POINTS FOR ANDERSON-DARLING STATISTIC
LOWER-TAIL PERCENTILE POINTS
5%: 0.1645341   10%: 0.1899085   15%: 0.2098461
UPPER-TAIL PERCENTILE POINTS
85%:   0.542403   90%:   0.610778   95%:   0.7264355   97.5%:
0.8407298   99%: 0.991330

For the transformed data, normality is not rejected for either the Anderson–Darling or the Bowman–Shenton test. This result suggests that the lognormal distribution is the appropriate probability distribution to model the data.

4

Factorial and Fractional Factorial Designs

An important problem in chemical research is that of finding effects of factors on responses of chemical processes. Chemical processes may depend on many factors, such as gas flow rates, detector conditions, pH, and chemical concentrations. Any one of these factors may affect analytical values, physical properties, or other responses that are of interest. A common practice is to investigate one factor while keeping other factors at a constant value and then to select another factor for the next set of experiments. However, this one-factor-at-a-time method has been shown to be inefficient, and it lacks the ability to detect interaction among factors.

An increase in efficiency can be gained by studying several factors simultaneously by using factorial experiments. These experiments can detect both factor effects and interactions among factors because every observation gives information about all factors.

4.1 Factorial Design Language

4.1.1 Factor Levels

Levels of a factor are labeled nominal, ordinal, interval, and ratio. *Nominal* levels cannot be ordered, nor is there a distance measurement between levels. Nominal levels would be used for such factors as analytical methods, laboratory technicians, or types of equipment. *Ordinal* levels have an ordered relationship but no measure of distance. Examples of ordinal levels are slow, medium, and fast, or smooth and rough. *Interval* levels have both an ordered relationship and a measure of distance but no measurement origin, such as zero. Meaningful comparisons between levels cannot be made without a measurement origin. For example, levels of 10°C and 20°C of a temperature factor can be ordered and a distance can be assigned between any two levels, but because 20°C is not twice as hot as 10°C, there is no measurement origin. Another example is levels of the factor pH, where pH = 4 is not twice the acidity of pH = 2. *Ratio* levels have an ordered relationship, a measure of distance, and a meaningful origin, such as levels for length, time, weight, and volume factors.

Qualitative factors have nominal levels, and quantitative factors have ordinal, interval, or ratio levels. Factorial experiments can be designed for both qualitative and quantitative factors. However, interpretation of results from factorial designs with qualitative factors must take precautions to account for the lack of an ordered relationship among factor levels. Factorial experiments designed for quantitative factors will be the main focus of this chapter.

4.1.2 Design Matrix

The number of factors in an experiment is K, and these factors will be represented by uppercase letters: $X_1, X_2, \ldots, X_K$. Levels of each factor will be represented by lowercase letters, x_{rh}, where the two subscripts represent the hth level of the rth factor. The number of levels may be different for each factor. For example, let factor X_1 have two levels and factor X_2 have three levels. The possible treatment combinations can be displayed in a matrix of rows and columns:

$$
\mathbf{D} \; = \;
\begin{bmatrix}
x_{11} & x_{21} \\
x_{11} & x_{22} \\
x_{11} & x_{23} \\
x_{12} & x_{21} \\
x_{12} & x_{22} \\
x_{12} & x_{23}
\end{bmatrix} .
$$

This matrix, where each row is a treatment combination, is called the *design matrix*. Note that $\mathbf{D}$ is boldface to denote an array of numbers or elements. The number of all possible treatment combinations (ie, rows) in a design matrix for an unreplicated factorial design is the product of all factor levels. For a replicated treatment combination, the corresponding row in the design matrix would be repeated. For example, suppose three factors have two levels, four factors have three levels, and each treatment combination is replicated five times. Then the number of design matrix rows is $N = 5 \times 2^3 \times 3^4 = 3240$. This example may be somewhat unusual, but experiments with factors having many levels will have a large number of treatment combinations. The number of design matrix rows is also the number of experimental units required to perform the experiment. Thus, there is a one-to-one correspondence between number of experimental units and number of rows in a design matrix.

Let factor X_1 be temperature, and let factor X_2 be pH with the levels

$$\text{Temperature (°C): 25, 50,}$$
$$\text{pH: 4, 7, 10.}$$

Then the design matrix for an unreplicated factorial design would be

$$\mathbf{D} = \begin{bmatrix} 25 & 4 \\ 25 & 7 \\ 25 & 10 \\ 50 & 4 \\ 50 & 7 \\ 50 & 10 \end{bmatrix}.$$

For clarity, treatment combination rows are usually ordered in a systematic manner in a design matrix. During the experiment, treatment combinations are run in a randomized order.

4.1.3 Model Matrix

For each treatment combination in the design matrix, a response, Y_i, $i = 1, 2, \ldots, N$, is observed. A factor effect is measured by response changes produced by changes in the factor's levels. Sometimes relationships between factors and a response can be expressed as a known mathematical function. When this relationship is unknown or not suggested by theoretical considerations, a polynomial model is often used to approximate the relationship. The simplest polynomial model involves only first-order or main effects. A first-order model for a two-factor example would express an observation by

$$Y_i = E(Y_i) + (\text{error})_i,$$
$$Y_i = B_o + B_1 X_{1i} + B_2 X_{2i} + (\text{error})_i,$$

where the second index on the factors corresponds to a row in the design matrix.

An observed random variable is the sum of its expected value and an error random variable. Further, the expected value is postulated to be modeled by a linear sum of a constant term, a first-order effect of factor X_1, and a first-order effect of factor X_2. Note that no assumptions have been made about the distributional properties of the error random variable. Later, when confidence intervals and hypothesis testing are investigated, distributional properties will be specified.

Coefficients B_o, B_1, and B_2 are model parameters that are unknown constants. An alternative way of expressing the expected value model is to use vector notation. A *vector* is either a column or a row of numbers or elements and, like a matrix, is boldface to indicate it is an array rather than a single number or element. Suppose for the ith observation, the corresponding row in the design matrix is the factor levels (x_{1h}, x_{2h}). The expected response is expressed in vector notation as

$$E(Y_i) = (1 \; x_{1h} \; x_{2h})_i \begin{bmatrix} B_0 \\ B_1 \\ B_2 \end{bmatrix}$$

$$= \mathbf{x}_i \mathbf{B}.$$

Vector multiplication is defined by multiplying the first element in row vector $\mathbf{x}_i$ by the first element in parameter column vector $\mathbf{B}$, plus the product of the second row element times the second column element, plus the third row element times the third column element. For vector multiplication the same number of elements must be in both row and column vectors.

Vector notation can be extended to represent all observations.

$$\begin{bmatrix} E(Y_1) \\ E(Y_2) \\ \cdot \\ \cdot \\ \cdot \\ E(Y_N) \end{bmatrix} = \begin{bmatrix} \mathbf{x}_1 \mathbf{B} \\ \mathbf{x}_2 \mathbf{B} \\ \cdot \\ \cdot \\ \cdot \\ \mathbf{x}_N \mathbf{B} \end{bmatrix} = \begin{bmatrix} \mathbf{x}_1 \\ \mathbf{x}_2 \\ \cdot \\ \cdot \\ \cdot \\ \mathbf{x}_N \end{bmatrix} \mathbf{B},$$

or

$$E(\mathbf{Y}) = \mathbf{XB}$$

This approximating model in matrix form succinctly represents the expected observation vector $E(\mathbf{Y})$ of size $N \times 1$ (N rows and 1 column) by the $N \times 3$ matrix $\mathbf{X}$ and the 3×1 parameter vector $\mathbf{B}$. The $\mathbf{X}$ matrix is called the *model matrix* and is derived from the design matrix by specifying the approximating model for $E(\mathbf{Y})$. For example, the first-order matrix model for the temperature and pH factors is

$$
E\begin{bmatrix} Y_1 \\ Y_2 \\ Y_3 \\ Y_4 \\ Y_5 \\ Y_6 \end{bmatrix} = \begin{bmatrix} 1 & 25 & 4 \\ 1 & 25 & 7 \\ 1 & 25 & 10 \\ 1 & 50 & 4 \\ 1 & 50 & 7 \\ 1 & 50 & 10 \end{bmatrix} \begin{bmatrix} B_0 \\ B_1 \\ B_2 \end{bmatrix}
$$

Matrix notation has the advantage of representing many different linear models with the same general form. The matrix solution for estimating parameters of any linear model is, therefore, one expression.

Suppose, from past experiments, the chemist knows that temperature effects depended on pH levels. This interaction between factors may be represented by including a second-order term in the linear model.

$$
E(Y_i) = B_0 + B_1 T_i + B_2 \text{pH}_i + B_{12}(T\text{xpH})_i .
$$

The second-order term $T\text{xpH}$ is easily included in the matrix model equation by expanding the $\mathbf{X}$ matrix by a column and the $\mathbf{B}$ vector by a row.

$$
E\begin{bmatrix} Y_1 \\ Y_2 \\ Y_3 \\ Y_4 \\ Y_5 \\ Y_6 \end{bmatrix} = \begin{bmatrix} 1 & 25 & 4 & 100 \\ 1 & 25 & 7 & 175 \\ 1 & 25 & 10 & 250 \\ 1 & 50 & 4 & 200 \\ 1 & 50 & 7 & 350 \\ 1 & 50 & 10 & 500 \end{bmatrix} \begin{bmatrix} B_0 \\ B_1 \\ B_2 \\ B_{12} \end{bmatrix}
$$

or

$$
E(\mathbf{Y}) = \mathbf{XB}.
$$

The additional interaction term increases the size of the $\mathbf{X}$ matrix to 6×4 and the size of the $\mathbf{B}$ matrix to 4×1.

To estimate the unknown parameter vector $\mathbf{B}$, matrix multiplication is used. Matrix multiplication is an extension of vector multiplication. The multiplication of $\mathbf{AB} = \mathbf{C}$ means each row vector in $\mathbf{A}$ multiplies each column vector in $\mathbf{B}$. There must be the same number of elements in row vectors of $\mathbf{A}$ as there are in column vectors of $\mathbf{B}$. This condition means that $\mathbf{A}$ is *conformable* with $\mathbf{B}$. If $\mathbf{A}$ is an $N \times K$ matrix (N rows and K columns) and $\mathbf{B}$ is a $K \times M$ matrix, their product is an $N \times M$ matrix $\mathbf{C}$. Although $\mathbf{A}$ can multiply $\mathbf{B}$, this does not necessarily mean that $\mathbf{B}$ can multiply $\mathbf{A}$.

For example, if **A** is a 3×2 matrix and **B** is a 2×2 matrix, then **C** = **AB** is a 3×2 matrix.

$$\mathbf{A} = \begin{bmatrix} 1 & 2 \\ 3 & 4 \\ 5 & 6 \end{bmatrix}, \quad \mathbf{B} = \begin{bmatrix} 7 & 8 \\ 9 & 10 \end{bmatrix}, \quad \mathbf{C} = \mathbf{AB} = \begin{bmatrix} 25 & 28 \\ 57 & 64 \\ 89 & 100 \end{bmatrix}.$$

The number in the first row and first column of **C** is the vector product of the first row vector of **A** and the first column vector of **B** (eg, $25 = 1 \times 7 + 2 \times 9$). Likewise, the ith row and jth column element in **C** is the product of the ith row vector of **A** and jth column vector of **B**. Note that the matrix multiplication **BA** is not defined because **B** is not conformable with **A**. However, if **A** is transposed, **A′**, then **B** is conformable with **A′**. A transposed matrix is indicated by a prime (′) and is an operation that changes column vectors into row vectors and row vectors into column vectors.

$$\mathbf{B} = \begin{bmatrix} 7 & 8 \\ 9 & 10 \end{bmatrix}, \quad \mathbf{A'} = \begin{bmatrix} 1 & 3 & 5 \\ 2 & 4 & 6 \end{bmatrix}, \quad \mathbf{C} = \mathbf{BA'} = \begin{bmatrix} 23 & 53 & 83 \\ 29 & 67 & 105 \end{bmatrix}.$$

For any $N \times K$ matrix **A**, the transposed matrix **A′** is always conformable with **A**, and the product **C** = **A′A** is a $K \times K$ square matrix.

A special matrix called the *identity matrix*, **I**, acts like the number 1 in matrix multiplication. The identity matrix is a square matrix with ones on the diagonal elements and zeros on the off diagonal elements. If **I** is an $N \times N$ identity matrix and **A** is an $N \times K$ matrix, then the product **IA** = **A**.

$$\mathbf{I} = \begin{bmatrix} 1 & 0 & 0 \\ 0 & 1 & 0 \\ 0 & 0 & 1 \end{bmatrix}, \quad \mathbf{A} = \begin{bmatrix} 1 & 2 \\ 3 & 4 \\ 5 & 6 \end{bmatrix}, \quad \mathbf{C} = \mathbf{IA} = \begin{bmatrix} 1 & 2 \\ 3 & 4 \\ 5 & 6 \end{bmatrix}.$$

4.1.4 Estimating the Parameter Vector

The purposes of many experiments are stated in terms of the unknown parameter vector **B**. The method of least squares will be used to estimate **B** by **b**. This method minimizes the sum of the squared residuals (ie, residual = response − estimated response):

$$\min \sum_{i=1}^{N} (Y_i - \hat{Y}_i)^2 = \min \mathbf{r'r},$$

where **r** is the $N \times 1$ residual vector in the fitted model:

$$Y = \hat{Y} + r = Xb + r.$$

Note that residuals, r, represent fitted errors, and the error terms added to the approximating model are represented by e. The solution for estimating the parameter vector is given by the matrix equation (Draper and Smith 1981)

$$b = (X'X)^{-1} X'Y.$$

Estimator b is composed of a matrix involving the model matrix and a vector of observed responses. An exponent of -1 represents the inverse of a matrix. The inverse of a square matrix (same number of rows and columns), A, has the property that $A^{-1}A = AA^{-1} = I$ where the matrix I is the identity matrix of ones on the diagonal elements and zeros on the off-diagonal elements. For two-level factorial and fractional factorial designs, the inverse of $X'X$ is easily calculated. Other designs may require the aid of a computer to calculate the inverse.

An important fact to remember is that b is just a linear combination of the observed responses. Because of this linear combination, the variance of b, obtained by using the assumptions that the errors are independent and $Var(e) = I\,\sigma^2$, is

$$Var(b) = (X'X)^{-1}\,\sigma^2.$$

This *variance-covariance matrix* has the variances of the parameter estimators down the diagonal and the covariances between estimators as off-diagonal elements.

4.1.5 Alias Matrix

The least squares estimator is an unbiased estimator, $E(b) = B$, provided the approximating model is the correct model. However, if the correct but unknown model has additional terms, the estimator using the approximating model will be biased. For example, let b_1 be the estimator for B_1 in the model

$$E(Y) = X_1 B_1,$$

but the true model is

$$E(Y) = X_1 B_1 + X_2 B_2.$$

The expected value of b_1 would be biased by additional parameters in B_2:

$$E(b_1) = B_1 + (X_1'X_1)^{-1} X_1'X_2 B_2.$$

Additional terms are *aliased* or *confounded* with parameters in the $\mathbf{B}_1$ vector. These terms are the extra terms that bias the estimator. The matrix multiplying the vector $\mathbf{B}_2$ is called the *alias matrix* and gives the structure of the aliasing. The alias matrix is the product of the model matrices of both approximating and true models and, therefore, depends on the design matrix. By selecting different design matrices, we can produce different aliasing structures. This fact will be used for selecting fractions of factorial designs.

Suppose a two-level factorial design is run. What effect would quadratic terms have on estimators for the intercept, main effects, and interactions? This effect will be illustrated for a 2^2 design using the alias matrix. Approximating model:

$$E(Y) = \begin{bmatrix} +1 & +1 & +1 & +1 \\ +1 & +1 & -1 & -1 \\ +1 & -1 & +1 & -1 \\ +1 & -1 & -1 & +1 \end{bmatrix} \begin{bmatrix} B_0 \\ B_1 \\ B_2 \\ B_{12} \end{bmatrix} = \mathbf{X}_1\mathbf{B}_1.$$

True model:

$$E(Y) = \begin{bmatrix} +1 & +1 & +1 & +1 \\ +1 & +1 & -1 & -1 \\ +1 & -1 & +1 & -1 \\ +1 & -1 & -1 & +1 \end{bmatrix} \begin{bmatrix} B_0 \\ B_1 \\ B_2 \\ B_{12} \end{bmatrix} + \begin{bmatrix} +1 & +1 \\ +1 & +1 \\ +1 & +1 \\ +1 & +1 \end{bmatrix} \begin{bmatrix} B_{11} \\ B_{22} \end{bmatrix}$$

$$= \mathbf{X}_1\mathbf{B}_1 + \mathbf{X}_2\mathbf{B}_2.$$

For this example an approximating model is assumed to have an intercept, two main effects, and an interaction term. However, if the true model also contains two quadratic terms, the expected value of the estimator for the model parameters is

$$E(\mathbf{b}) = \mathbf{B}_1 + (\mathbf{X}_1'\mathbf{X}_1)^{-1}\mathbf{X}_1'\mathbf{X}_2\mathbf{B}_2$$

$$= \mathbf{B}_1 + \frac{1}{4}\mathbf{I} \begin{bmatrix} 4 & 4 \\ 0 & 0 \\ 0 & 0 \\ 0 & 0 \end{bmatrix} \begin{bmatrix} B_{11} \\ B_{22} \end{bmatrix}$$

$$= \begin{bmatrix} B_0 + B_{11} + B_{22} \\ B_1 \\ B_2 \\ B_{12} \end{bmatrix}$$

This analysis shows that the intercept is aliased or confounded with two quadratic coefficients B_{11} and B_{22}, and the estimators for B_1, B_2, and B_{12} are unbiased. Therefore, when quadratic effects are present, least squares estimators of main effects and interactions are unbiased for a two-level factorial design, and the intercept estimator is biased.

When many factors are being examined, two-level factorials can be used to examine which main effects cause significant effects on the response even in the presence of quadratic effects. Those factors that do not cause significant effects can be eliminated, and a three-level factorial can be performed to study quadratic effects by using a smaller number of factors.

4.2 2^K Factorial Designs

The most common types of factorial designs are those that have all factors at two levels (eg, low and high levels). The number of different treatment combinations is 2^K , and if r replicates are performed at each treatment combination, the number of experimental units is $N = r2^K$. To study two-level factorial designs, the levels will be coded -1 (low level) and $+1$ (high level), so that treatment combinations are represented by a series of plus and minus signs [eg, $(-,+,-)$ means (low,high,low)]. For an air pressure factor with levels 20 and 42 psig, the levels can be coded by

$$\text{low level} = \frac{2(\text{low value} - \text{mean})}{\text{range}} = \frac{2(20 - 31)}{22} = -1,$$

$$\text{high level} = \frac{2(\text{high value} - \text{mean})}{\text{range}} = \frac{2(42 - 31)}{22} = +1.$$

The design matrix of a 2^K factorial design is a listing of the total number of treatment combinations. A 2^3 factorial design would have $N = 8$ treatment combinations with design matrix

$$\mathbf{D} = \begin{bmatrix} + & + & + \\ + & + & - \\ + & - & + \\ + & - & - \\ - & + & + \\ - & + & - \\ - & - & + \\ - & - & - \end{bmatrix}$$

A 2^K factorial model matrix includes the mean, first-order effects, and all two-factor, three-factor, . . . , K-factor interactions. An *interaction* is a measure of the effect on the response, due to changing the levels of one factor, that depends on the levels of one or more other factors. When factors are not independent, the data requires detailed study. If numbers **1, 2, 3** represent column vectors in **X** corresponding to the three first-order or main effects, the model matrix for a 2^3 factorial design is

$$
\mathbf{X} = \begin{array}{c} \mathbf{I} \quad \mathbf{1} \quad \mathbf{2} \quad \mathbf{3} \quad \mathbf{12} \quad \mathbf{13} \quad \mathbf{23} \quad \mathbf{123} \\ \left[\begin{array}{cccccccc}
+ & + & + & + & + & + & + & + \\
+ & + & + & - & + & - & - & - \\
+ & + & - & + & - & + & - & - \\
+ & + & - & - & - & - & + & + \\
+ & - & + & + & - & - & + & - \\
+ & - & + & - & - & + & - & + \\
+ & - & - & + & + & - & - & + \\
+ & - & - & - & + & + & + & -
\end{array}\right]
\end{array}
$$

The corresponding parameter vector is $\mathbf{B}' = (B_0, B_1, B_2, B_3, B_{12}, B_{13}, B_{23}, B_{123})$. The first column in the model matrix will be the multiplier for B_0; the second column will be the multiplier for B_1, and so on. The model matrix is formed by making the first column all $+1$. The columns for **1, 2, 3** are the same as the design matrix. Columns for two-factor interactions **1∗2, 1∗3**, and **2∗3** are formed by multiplying elements in the rows of the two corresponding columns in the design matrix. Three-factor interaction, **1∗2∗3**, is formed by multiplying the elements in the rows of all three columns in the design matrix.

To illustrate the matrix equation for estimating parameter vectors, let $\mathbf{Y}' = (Y_1, Y_2, Y_3, Y_4, Y_5, Y_6, Y_7, Y_8)$ be the ordered response measurements for 2^3 factorial treatment combinations. The least squares estimator of the parameter vector starts by multiplying $\mathbf{X}'\mathbf{X}$:

$$
\mathbf{X}'\mathbf{X} = \begin{bmatrix}
8 & 0 & 0 & 0 & 0 & 0 & 0 & 0 \\
0 & 8 & 0 & 0 & 0 & 0 & 0 & 0 \\
0 & 0 & 8 & 0 & 0 & 0 & 0 & 0 \\
0 & 0 & 0 & 8 & 0 & 0 & 0 & 0 \\
0 & 0 & 0 & 0 & 8 & 0 & 0 & 0 \\
0 & 0 & 0 & 0 & 0 & 8 & 0 & 0 \\
0 & 0 & 0 & 0 & 0 & 0 & 8 & 0 \\
0 & 0 & 0 & 0 & 0 & 0 & 0 & 8
\end{bmatrix}
$$

The first row vector of $\mathbf{X}'$ multiplies the first column vector of $\mathbf{X}$. This multiplication is equivalent to squaring and then summing all the entries in the first column of $\mathbf{X}$. Because all values are ± 1, the value is 8. The product of the first row vector of $\mathbf{X}'$ with any other column vector of $\mathbf{X}$ has the value zero. If the vector product $\mathbf{c}'\mathbf{d} = 0$, then vector $\mathbf{c}$ is *orthogonal* to vector $\mathbf{d}$. The same pattern holds for the remaining matrix multiplication, giving the results 8 on the diagonals and 0 on the off-diagonals, or

$$\mathbf{X}'\mathbf{X} = 8 \begin{bmatrix} 1 & 0 & 0 & 0 & 0 & 0 & 0 & 0 \\ 0 & 1 & 0 & 0 & 0 & 0 & 0 & 0 \\ 0 & 0 & 1 & 0 & 0 & 0 & 0 & 0 \\ 0 & 0 & 0 & 1 & 0 & 0 & 0 & 0 \\ 0 & 0 & 0 & 0 & 1 & 0 & 0 & 0 \\ 0 & 0 & 0 & 0 & 0 & 1 & 0 & 0 \\ 0 & 0 & 0 & 0 & 0 & 0 & 1 & 0 \\ 0 & 0 & 0 & 0 & 0 & 0 & 0 & 1 \end{bmatrix} = 8\mathbf{I}.$$

In general, a design is called an *orthogonal design* with respect to the approximating model if the $\mathbf{X}'\mathbf{X}$ matrix has only non-zero diagonal elements.

An advantage of using 2^K factorial designs is that these designs have the orthogonality property. This property simplifies the inverse calculation of the $\mathbf{X}'\mathbf{X}$ matrix:

$$(\mathbf{X}'\mathbf{X})^{-1} = \frac{1}{8}\mathbf{I}.$$

The estimated parameter vector is

$$\mathbf{b} = \frac{1}{8}\mathbf{I}\mathbf{X}'\mathbf{Y} = \frac{1}{8}\mathbf{X}'\mathbf{Y}.$$

$$b_0 = \frac{1}{8}(+Y_1 + Y_2 + Y_3 + Y_4 + Y_5 + Y_6 + Y_7 + Y_8),$$

$$b_1 = \frac{1}{8}(+Y_1 + Y_2 + Y_3 + Y_4 - Y_5 - Y_6 - Y_7 - Y_8),$$

$$b_2 = \frac{1}{8}(+Y_1 + Y_2 - Y_3 - Y_4 + Y_5 + Y_6 - Y_7 - Y_8),$$

$$b_3 = \frac{1}{8}(+Y_1 - Y_2 + Y_3 - Y_4 + Y_5 - Y_6 + Y_7 - Y_8),$$

$$b_{12} = \frac{1}{8}(+Y_1 + Y_2 - Y_3 - Y_4 - Y_5 - Y_6 + Y_7 + Y_8),$$

$$b_{13} = \frac{1}{8}(+Y_1 - Y_2 + Y_3 - Y_4 - Y_5 + Y_6 - Y_7 + Y_8),$$

$$b_{23} = \frac{1}{8}(+Y_1 - Y_2 - Y_3 + Y_4 + Y_5 - Y_6 - Y_7 + Y_8),$$

$$b_{123} = \frac{1}{8}(+Y_1 - Y_2 - Y_3 + Y_4 - Y_5 + Y_6 + Y_7 - Y_8).$$

Parameter estimators are linear combinations of observed values, with b_0 equal to the average of the responses. Estimated parameters for main effects b_1, b_2, and b_3 are equal to the average of the difference between responses at the high level and low level for the corresponding factor.

Variances of linear combinations may be calculated by methods in Chapter 3 or obtained from diagonal elements of the variance-covariance matrix. The variance-covariance matrix consists of variances of parameter estimators and covariances between these estimators:

$$\text{Var}(\mathbf{b}) = (\mathbf{X'X})^{-1}\sigma^2 = \frac{\sigma^2}{8}\mathbf{I}.$$

The variance of each parameter estimator is on a corresponding diagonal element, and the covariance between any two estimators is on a corresponding row and column off-diagonal element. Estimated parameter variances for full two-level factorial designs are all equal, and covariances between estimators are zero. For this 2^3 example, the variance of any estimator is $\sigma^2/8$. An estimate of σ^2 can be made from this experiment if either replicate experimental units are run for some or all treatment combinations or interactions are assumed to be zero. Occasionally, σ^2 is available from a previous experiment, or can it be found in the literature. Another practice is to assume, before the experiment, that the interaction terms are zero and to use only a first-order model to fit the data. Additional information previously used to estimate interaction parameters can now be used to estimate the variance σ^2. This method should only be used if we are confident that the first-order model is correct.

Confidence intervals on parameters of a 2^K factorial design can be calculated if the additional assumption is made that the errors are i.i.d $N(0, \sigma^2)$. If the model is correct, an unbiased estimator of σ^2 can be calculated from the sum of the squared residuals of the fitted model by using S^2 with df degrees of freedom, where

$$S^2 = \sum_{i=1}^{N} \frac{(Y_i - \hat{Y}_i)^2}{\text{df}}$$

(ie, df = number of observations − number of model parameters). For parameter B_j the $100(1 - \alpha)\%$ confidence interval is

$$b_j \pm t_{\alpha/2}(\text{df}) \sqrt{\frac{S^2}{2^K}}.$$

If this interval includes zero, the data indicates that the parameter B_j is not significantly different from zero at the $\alpha \times 100\%$ significance level.

4.3 3^K Factorial Designs

Second-order effects in a 2^K factorial design are restricted to two-factor interactions, but quadratic curvature of a response cannot be estimated using only two levels. If a factor effect causes a response to increase and then decrease, or vice versa, more than two levels are required to estimate this effect. Three-level factorial designs can be used to measure curvature. Three equally spaced levels for each factor are first coded to values of $-1, 0, +1$:

$$\text{low level} \; = \; \frac{2(\text{low value} - \text{average})}{\text{range}} = -1,$$

$$\text{middle level} = \frac{2(\text{middle value} - \text{average})}{\text{range}} = 0,$$

$$\text{high level} \;\; = \; \frac{2(\text{high value} - \text{average})}{\text{range}} = +1.$$

The design matrix for a 3^2 factorial design would be

$$\mathbf{D} = \begin{bmatrix} +1 & +1 \\ +1 & 0 \\ +1 & -1 \\ 0 & +1 \\ 0 & 0 \\ 0 & -1 \\ -1 & +1 \\ -1 & 0 \\ -1 & -1 \end{bmatrix}$$

Responses can be represented by a second-order model that includes not only mixed terms between the two factors but also the squared terms of each factor.

$$E(Y_i) = B_0 + B_1 X_{1i} + B_2 X_{2i} + B_{11} X_{1i}^{\,2} + B_{22} X_{2i}^{2} + B_{12} X_{1i} X_{2i}.$$

Note that the index on each second-order coefficient is a double subscript which indicates the corresponding factors. The model matrix would be

$$
\mathbf{X} = \begin{array}{cccccc}
\mathbf{I} & \mathbf{1} & \mathbf{2} & \mathbf{11} & \mathbf{22} & \mathbf{12} \\
\left[\begin{array}{cccccc}
+1 & +1 & +1 & +1 & +1 & +1 \\
+1 & +1 & 0 & +1 & 0 & 0 \\
+1 & +1 & -1 & +1 & +1 & -1 \\
+1 & 0 & +1 & 0 & +1 & 0 \\
+1 & 0 & 0 & 0 & 0 & 0 \\
+1 & 0 & -1 & 0 & +1 & 0 \\
+1 & -1 & +1 & +1 & +1 & -1 \\
+1 & -1 & 0 & +1 & 0 & 0 \\
+1 & -1 & -1 & +1 & +1 & +1
\end{array}\right]
\end{array},
$$

with

$$
\mathbf{X'X} = \begin{bmatrix}
9 & 0 & 0 & 6 & 6 & 0 \\
0 & 6 & 0 & 0 & 0 & 0 \\
0 & 0 & 6 & 0 & 0 & 0 \\
6 & 0 & 0 & 6 & 4 & 0 \\
6 & 0 & 0 & 4 & 6 & 0 \\
0 & 0 & 0 & 0 & 0 & 4
\end{bmatrix}.
$$

For this 3^2 factorial design, row vectors in $\mathbf{X'}$ are not necessarily orthogonal to column vectors in $\mathbf{X}$. As a result there is no simple formula for calculating the inverse of $\mathbf{X'X}$, and either linear algebra methods (Graybill 1969) or a computer program must be used.

$$
(\mathbf{X'X})^{-1} = \begin{bmatrix}
\frac{5}{9} & 0 & 0 & -\frac{1}{3} & -\frac{1}{3} & 0 \\
0 & \frac{1}{6} & 0 & 0 & 0 & 0 \\
0 & 0 & \frac{1}{6} & 0 & 0 & 0 \\
-\frac{1}{3} & 0 & 0 & \frac{1}{2} & 0 & 0 \\
-\frac{1}{3} & 0 & 0 & 0 & \frac{1}{2} & 0 \\
0 & 0 & 0 & 0 & 0 & \frac{1}{4}
\end{bmatrix}.
$$

Using this inverse, we obtain the parameter estimators for the 3^2:

$\mathbf{b} = (\mathbf{X'X})^{-1}\mathbf{X'Y}$:

$$b_0 = \frac{1}{9}(-Y_1 + 2Y_2 - Y_3 + 2Y_4 + 5Y_5 + 2Y_6 - Y_7 + 2Y_8 - Y_9),$$

$$b_1 = \frac{1}{6}(Y_1 + Y_2 + Y_3 - Y_7 - Y_8 - Y_9),$$

$$b_2 = \frac{1}{6}(Y_1 - Y_3 + Y_4 - Y_6 + Y_7 - Y_9),$$

$$b_{11} = \frac{1}{6}(Y_1 + Y_2 + Y_3 - 2Y_4 - 2Y_5 - 2Y_6 + Y_7 + Y_8 + Y_9),$$

$$b_{22} = \frac{1}{6}(Y_1 - 2Y_2 + Y_3 + Y_4 - 2Y_5 + Y_6 + Y_7 - 2Y_8 + Y_9),$$

$$b_{12} = \frac{1}{4}(Y_1 - Y_3 - Y_7 + Y_9).$$

The estimator of the intercept, b_0, is a linear combination of all responses, whereas the estimators of linear coefficients, b_1 and b_2, are the average of only the difference between the responses at high and low levels of their respective factors X_1 and X_2. Estimators of pure quadratic coefficients, b_{11} and b_{22}, are also a linear combination of all responses. Estimator b_{12} is the average of the difference between responses when both factors are at either the high level or the low level and responses when one factor is at the high (low) level while the other is at the low (high) level.

Variance estimates of coefficient estimators require an estimate of σ^2 for the experimental error from either an independent source or replicated experimental units. However, by assuming the model is correct, an unbiased variance estimator for the experimental error is given by

$$S^2 = \sum_{i=1}^{N} (Y_i - \hat{Y}_i)^2/\mathrm{df}$$

(ie, df = number of responses − number of model parameters). Variance and covariance estimates of coefficient estimators are then calculated by $\widehat{\mathrm{Var}}(\mathbf{b}) = (\mathbf{X'X})^{-1} S^2$. Diagonal elements of this matrix are the variance estimates, and off-diagonal elements are covariance estimates. The first diagonal element is $\widehat{\mathrm{Var}}(b_0)$, the second diagonal element is $\widehat{\mathrm{Var}}(b_1)$, and so on.

$$\widehat{\mathrm{Var}}(b_0) = \frac{5S^2}{9}, \qquad \widehat{\mathrm{Var}}(b_1) = \widehat{\mathrm{Var}}(b_2) = \frac{S^2}{6},$$

$$\widehat{\mathrm{Var}}(b_{11}) = \widehat{\mathrm{Var}}(b_{22}) = \frac{S^2}{2}, \qquad \text{and} \qquad \widehat{\mathrm{Var}}(b_{12}) = \frac{S^2}{4}.$$

Nonzero off-diagonal elements of $\widehat{\text{Var}}(\mathbf{b})$ are estimates of covariances between the corresponding coefficient estimators. Nonzero estimated covariances occur in the first row at the fourth and fifth columns (or in the first column at the fourth and fifth rows because the matrix is symmetrical):

$$\widehat{\text{Cov}}(b_0, b_1) = \widehat{\text{Cov}}(b_0, b_2) = -\frac{S^2}{3}.$$

Using the variance estimates, we can calculate confidence intervals for any coefficient value, B, by

$$\hat{B} \pm (t\text{-value}) \sqrt{\widehat{\text{Var}}(\hat{B})}.$$

If this interval contains zero, the value of the coefficient is not significantly different from zero at the desired significance level. Similarly, confidence intervals for an estimated expected response at a point $\mathbf{x}_h' = (1 \; x_{1h} \; x_{2h} \; x_{1h}^2 \; x_{2h}^2 \; x_{1h}x_{2h})$ can be calculated once the estimated variance is calculated. The expected response estimator is a linear combination of random variables (ie, coefficient estimators); therefore

$$\hat{E}(Y_h) = \hat{Y}_h = \mathbf{x}_h'\mathbf{b},$$

$$\text{Var}(\hat{Y}_h) = \mathbf{x}_h'\text{Var}(\mathbf{b})\mathbf{x}_h,$$

or

$$\widehat{\text{Var}}(\hat{Y}_h) = S^2 \left(\frac{5}{9} - \frac{(x_{1h}^2 + x_{2h}^2 - x_{1h}^4 - x_{2h}^4)}{2} + \frac{(x_{1h}x_{2h})^2}{4} \right).$$

The variance of expected value estimators depends on (x_{1h}, x_{2h}) or their location in the experimental region. At the design corners [ie, $(\pm 1, \pm 1)$], $\widehat{\text{Var}}(\hat{Y}) = {}^{29}/_{36} \, S^2$ and at the remaining design points $(0,0)$, $(0, \pm 1)$, and $(\pm 1, 0)$, the estimated variance value is $\widehat{\text{Var}}(\hat{Y}) = {}^{5}/_{9} \, S^2$. Variances of expected response estimators may be calculated for any other location in the experimental region's interior.

Because variances of response estimators depend on the location of design points, the experimenter now has a criterion for selecting experimental designs. For example, chemists may want the variance of the expected response to have equal values at equal distances from the design's center. This criterion is known as *rotatability*, and experimental designs having this property are known as *rotatable designs*. Other criteria may also be desired, such as weighting the importance of some response variances more than others. Because the variance of an estimated response can be calculated at

any design point, confidence intervals on an expected response follow the
usual formula

$$\hat{Y}_h \pm t \text{ (df) } \sqrt{\hat{\text{Var}}(\hat{Y}_h)},$$

where the t-value is selected at the desired confidence coefficients and the
degrees of freedom, df, are used to calculate S^2. Another design criterion,
D-optimality, is related to minimizing the volume of simultaneous confidence intervals for parameter values. Design points are found by selecting
those points that minimize the determinant of the matrix $\mathbf{X'X}$.

In choosing between two-level and three-level factorial designs, chemists must weigh the properties of each design. Two-level designs are highly
efficient designs that can estimate the main effects and interactions.
Three-level designs use more design points but can estimate main effects,
interactions, and pure quadratic effects. Quadratic effects can be very
important, and unless previous experience is available on the response's
behavior, these effects should be investigated.

4.4 Fractional Factorial Designs

A disadvantage of a full-factorial design is that the number of experimental units required increases rapidly as the number of factors increases.
For example, two-level factorial designs with $K = 2, 3, 4, 5, 6,$ and 7
require $N = 4, 8, 16, 32, 64,$ and 128 experimental units, respectively, to
estimate all parameters in a full-factorial model. Chemists frequently feel
that running a full-factorial model for many factors ($K > 3$) is either too
costly or requires too much time. One solution is to hold some of the factors
at a constant level and run a smaller design. However, this solution restricts
the experimental scope and is not necessary if information on higher
interactions can be sacrificed. For example, the number of parameters for
intercept, main effects, and higher interactions for 2^K full-factorial models
are given in the following table, where FI = factor interaction.

Parameters	$K=2$	$K=3$	$K=4$	$K=5$	$K=6$	$K=7$
Intercept	1	1	1	1	1	1
Main	2	3	4	5	6	7
2 FI	1	3	6	10	15	21
3 FI		1	4	10	20	35
4 FI			1	5	15	35
5 FI				1	6	21
6 FI					1	7
7 FI						1

If the objective of an experiment is to study main effects of factors, a fraction of the full-factorial design can be used. This fractional factorial design should allow estimation of main effects, so that their expected values are aliased to only higher-order interactions. Higher-order interactions are effects that are most likely to be nonsignificant, if they exist. For example, a 2^4 full-factorial design can be used to estimate the parameters

Intercept	Main	2 FI	3 FI	4 FI
B_0	B_1	B_{12}	B_{123}	B_{1234}
	B_2	B_{13}	B_{124}	
	B_3	B_{14}	B_{134}	
	B_4	B_{23}	B_{234}	
		B_{24}		
		B_{34}		

Suppose a half-fraction of the full 2^4 factorial (one-half of 2^4, or 2^{4-1}) can be designed so that the following parameters can be estimated:

Intercept	Main Effect	2 FI Effect
$B_0 + B_{1234}$	$B_1 + B_{234}$	$B_{12} + B_{34}$
	$B_2 + B_{134}$	$B_{13} + B_{24}$
	$B_3 + B_{124}$	$B_{14} + B_{23}$
	$B_4 + B_{123}$	

All these parameters are aliased or confounded with other parameters. The alias structure, however, is chosen so that the highest-order interaction is confounded with the mean, the next highest with the main effects, and the two-factor interactions are confounded with each other. If three- and four-factor interaction parameters are negligible, the intercept and main-effect estimators would be unbiased.

For two parameters to be aliased means either that their corresponding columns in the model matrix are identical, or that the column corresponding to one effect is the negative of the column corresponding to the other. Let **1**, **2**, **3**, and **4** represent columns corresponding to main-effect parameters B_1, B_2, B_3, and B_4. Using this notation, we can write the column corresponding to B_{1234} as **1*2*3*4**, which represents multiplication of elements in the rows of the four columns. This is not the same as vector multiplication. The column vector corresponding to the intercept is composed of all ones and is represented by I. For B_0 to be aliased with B_{1234} means

$$I = 1*2*3*4 \quad \text{or} \quad -I = 1*2*3*4.$$

This relationship is called the *defining relation* or *defining contrast*. The defining relation for a half-fraction of a 2^K is

$$\pm I = 1*2* \cdots *K.$$

The product across row elements may be either positive or negative, because B_0 can be confounded to $+B_{12} \cdots {}_K$ or $-B_{12} \cdots {}_K$.

The defining relation is used to construct half-fractions of a 2^K factorial by the following rules:

Rule 1. Write down a full 2^{K-1} factorial.

Example. To construct a half-fraction of 2^4, start with a complete 2^3 factorial design.

1	2	3
+	+	+
+	+	−
+	−	+
+	−	−
−	+	+
−	+	−
−	−	+
−	−	−

Rule 2. Write down the Kth column so that the defining relation $\pm I = 1*2* \cdots *K$ is satisfied.

Example. $+I = 1*2*3*4.$

1	2	3	4
+	+	+	+
+	+	−	−
+	−	+	−
+	−	−	+
−	+	+	−
−	+	−	+
−	−	+	+
−	−	−	−

The product is one ($+1$) if elements across each row are multiplied together. This multiplication produces a column of ones for the B_{1234} column vector and satisfies the defining relation.

Rule 3. Determine which parameters are confounded with a selected parameter by multiplying the corresponding design column vector times both sides of the defining relation. Multiplication of I leaves any column vector unchanged. Multiplication of any column vector by itself squares all the elements in the column vector. A squared column vector of positive ones is equal to I, so that the relationship, $I = 1^2 = 2^2 = \cdots = K^2$, is used for this multiplication.

  Example. Determine the parameters confounded with B_{13} for the defining relation $I = 1*2*3*4$.

$$1*3*I = 1^2*2*3^2*4 = 2*4.$$

Therefore, B_{13} is confounded with B_{24}. This means that if b_{13} is the least squares estimator of B_{13}, then $E(b_{13}) = B_{13} + B_{24}$.

  Analysis of estimated parameters from fractional factorial designs can be used to screen out those factors that are not significant. Subsequent experiments with a smaller number of factors can then be run to investigate interactions and curvature. Fractions of factorial designs can also be used when there is a need to block a factorial design. Suppose a 2^4 design is run on two different days, and a day effect on the response is expected due to the time lapse. This day effect can be accounted for by running a 2^{4-1} design by using the defining relation $+I = 1*2*3*4$ one day, and the complentary fraction by using $-I = 1*2*3*4$ the second day. Expected values of main-effect estimators for the two fractions, say b_1 and b_1*, are

$$\begin{aligned} \text{Day 1:} \quad E(b_1) &= B_1 + B_{234}, \\ \text{Day 2:} \quad E(b_1*) &= B_1 - B_{234}. \end{aligned}$$

Unbiased estimators of B_1 and B_{234} are obtained by averaging the sum and difference of estimators from the two fractions.

$$1/2\ E(b_1 + b_1*) = B_1, \quad 1/2E(b_1 - b_1*) = B_{234}.$$

Unbiased estimators of other coefficients are similarly obtained.

  A half-fraction of a factorial may still require too many experimental units, given budget and time constraints. For example, when an experiment centers on only five main effects or first-order parameters, a 2^{5-1} design requiring 16 experimental units may not be cost effective. For this situation a quarter-fraction, 2^{5-2}, can be designed to estimate main effects in only eight runs.

Construction of a quarter-fraction requires two defining relations. In addition, the product of the two defining relations will also be aliased to **I**. The defining relation must be chosen carefully, so that unwanted confounding does not occur. For example, suppose the following defining relation is chosen:

$$\mathbf{I} = +1*2*3*4*5 = -1*2*3 = -4*5.$$

The first two defining relations are generators, and the last defining relation is the product of generators. This defining relation is a poor choice because first-order coefficients B_4 and B_5 are confounded.

$$4*\mathbf{I} = +1*2*3*4^2*5 = -1*2*3*4 = -4^2*5,$$
$$4 = +1*2*3*5 = -1*2*3*4 = -5.$$

A construction method for quarter-fractions of a 2^K factorial is as follows:

Rule 1. Write out a full 2^{K-2} factorial design.

   Example. To construct a quarter-fraction of 2^5 (ie, 2^{5-2}), start with a complete 2^3 factorial design.

1	2	3
+	+	+
+	+	−
+	−	+
+	−	−
−	+	+
−	+	−
−	−	+
−	−	−

Rule 2. Write out the columns for two of the highest-order interactions of the $K-2$ factors to represent the remaining two factors. Multiply each interaction column either by $+1$ or -1, choosing the multiplier by a random process (ie, flip of a fair coin).

Example.

1	2	3	$+1*2*3$	$-1*2$
+	+	+	+	−
+	+	−	−	−
+	−	+	−	+
+	−	−	+	+
−	+	+	−	+
−	+	−	+	+
−	−	+	+	−
−	−	−	−	−

Rule 3. Generators are found by multiplying the new column vectors (ie, **4** and **5**) by the generating interactions (ie, $\mathbf{I} = \mathbf{1*2*3*4} = \mathbf{-1*2*5}$). The defining relation consists of the two generators and their product. The alias structure is found by multiplying the defining relation by the appropriate column vector. Like rule 3 for half-fractions, the relationship $\mathbf{I} = \mathbf{1}^2 = \mathbf{2}^2 = \cdots = \mathbf{K}^2$ is used for multiplying column vectors.

Example.

$$
\begin{aligned}
\mathbf{I} &= \mathbf{1*2*3*4} & &= \mathbf{-1*2*5} & &= \mathbf{-3*4*5} \\
\mathbf{1} &= \mathbf{2*3*4} & &= \mathbf{-2*5} & &= \mathbf{-1*3*4*5} \\
\mathbf{2} &= \mathbf{1*3*4} & &= \mathbf{-1*5} & &= \mathbf{-2*3*4*5} \\
\mathbf{3} &= \mathbf{1*2*4} & &= \mathbf{-1*2*3*5} & &= \mathbf{-4*5} \\
\mathbf{4} &= \mathbf{1*2*3} & &= \mathbf{-1*2*4*5} & &= \mathbf{-3*5} \\
\mathbf{5} &= \mathbf{1*2*3*4*5} & &= \mathbf{-1*2} & &= \mathbf{-3*4} \\
\mathbf{1*3} &= \mathbf{2*4} & &= \mathbf{-2*3*5} & &= \mathbf{-1*4*5} \\
\mathbf{1*4} &= \mathbf{2*3} & &= \mathbf{-2*4*5} & &= \mathbf{-1*3*5}
\end{aligned}
$$

For the above 2^{5-2} fractional factorial, the estimators of the first-order effects are unbiased only if two-factor and higher interaction effects are assumed to be negligible or zero. Fractional factorial designs are classified by the order of assumed negligible effects. A fractional factorial design is an even or odd *resolution* (John 1971) according to the following table:

Resolution	Order of estimable effects	Order of negligible effects
$2t + 1$	t or less	$t + 1$ or greater
$2t$	$t - 1$ or less	$t + 1$ or greater

The quarter-fraction defined by $I = 1*2*3*4 = -1*2*5 = -3*4*5$ is a resolution III design (roman numerals are used to designate resolution). For fractional factorials constructed by methods in this section, resolution of a design is the smallest number of columns used to represent an interaction in the defining relation. For the half-fraction of 2^5 defined by $I = 1*2*3*4*5$, the design is of resolution V, which means all first- and second-order effects can be estimated if third- and higher-order effects are negligible or zero.

The minimum number of experimental units for a resolution III design for K-factors is $N = K + 1$. The notation $2^K //N$ will be used to denote an N-point fraction of a 2^K factorial. Can resolution III designs for $2^K //K + 1$ be found for all values of K? The answer is yes; but not all *saturated designs*, those designs that have only one more experimental unit than the number of factors (ie, $2^K //K + 1$ designs), are orthogonal. Examples of orthogonal designs are the half-fraction 2^{3-1} and the 16th fraction 2^{7-4}, which is constructed in the same manner as a quarter-fraction. In fact, whenever $K + 1$ is a power of two, a saturated design of the form 2^{K-P} can be constructed for the proper value of P.

Another important class of resolution III saturated designs is Plackett and Burman designs. *Plackett and Burman designs* are orthogonal saturated designs for cases when $K + 1$ is any multiple of four. The design matrices for $K + 1 = 4$ to 100 (except for $K + 1 = 92$) are listed in Plackett and Burman (1946). Actually, only the first rows of the design matrices are given, from which the remaining designs can be constructed. First rows for Plackett and Burman designs for $N = 8$, 12, 16, 20, and 24 are listed in Table 4.1.

TABLE 4.1 First row of Plackett and Burman designs

K	N	First row of the design matrix
7	8	+ + + − + − −
11	12	+ + − + + + − − − + −
15	16	+ + + + − + − + + − − + − − −
19	20	+ + − − + + + + − + − + − − − − + + −
23	24	+ + + + + − + − + + − − + + − − + − + − − − −

The complete design matrix is generated by shifting the first row cyclically and adding a final row of minus signs. Shifting is done by moving the element in the last column to the element in the first column of the next row and then shifting the remaining elements over one column.

Example. $2^7 //8$.

$$
\mathbf{D} = \begin{bmatrix}
+ & + & + & - & + & - & - \\
- & + & + & + & - & + & - \\
- & - & + & + & + & - & + \\
+ & - & - & + & + & + & - \\
- & + & - & - & + & + & + \\
+ & - & + & - & - & + & + \\
+ & + & - & + & - & - & + \\
- & - & - & - & - & - & -
\end{bmatrix} .
$$

The response would be represented by a first-order model with an intercept and main effects only. The model matrix is constructed by adding an initial column of ones to $\mathbf{D}$. The model matrix is orthogonal and therefore $\mathbf{X'X} = (K+1)\mathbf{I}$. Coefficient estimators and their variances are given by

$$
\mathbf{b} = \frac{1}{K+1}\mathbf{X'Y} \qquad \text{and} \qquad \text{Var}(\mathbf{b}) = \frac{\sigma^2}{K+1}\mathbf{I} .
$$

The coefficient estimators are uncorrelated with each other, and all have the same variance.

Orthogonal designs can also be derived from Plackett and Burman designs when the number of factors does not meet the requirement that $K+1$ is a multiple of four. These designs will not be saturated but are resolution III designs constructed with a small number of experimental units. For example, an orthogonal resolution III design can be constructed for a $2^5//8$ by using the Plackett and Burman design $2^7//8$. The $2^5//8$ design is constructed by selecting any of the five columns (usually a random selection is preferable) in the Plackett and Burman design. Because the original design is orthogonal, any five columns will be orthogonal. Suppose the first five columns are selected; the design matrix for the $2^5//8$ is given by

$$
\mathbf{D} = \begin{bmatrix}
+ & + & + & - & + \\
- & + & + & + & - \\
- & - & + & + & + \\
+ & - & - & + & + \\
- & + & - & - & + \\
+ & - & + & - & - \\
+ & + & - & + & - \\
- & - & - & - & -
\end{bmatrix} .
$$

Therefore, orthogonal resolution III designs for K factors can be constructed by using a Plackett and Burman design in $K+1$ runs or the next largest design. If $K+1$ is a multiple of four, the resolution III design will be saturated.

The alias structure of resolution III designs constructed from Plackett and Burman designs is not easy to calculate because these designs are not constructed from defining relations. The alias structure is found by using the general method described in the subsection "Alias Matrix." For example, let the true model for a seven-factor experiment include the intercept, main effects, and all two-factor interactions. If a $2^7//8$ Plackett and Burman design is performed, assuming a first-order model, the following parameters are aliased:

$$E(b_0) = B_0, \qquad\qquad E(b_4) = B_4 - B_{13} - B_{25} - B_{67},$$
$$E(b_1) = B_1 - B_{26} - B_{34} - B_{57}, \qquad E(b_5) = B_5 - B_{17} - B_{24} - B_{36},$$
$$E(b_2) = B_2 - B_{16} - B_{37} - B_{45}, \qquad E(b_6) = B_6 - B_{12} - B_{35} - B_{47},$$
$$E(b_3) = B_3 - B_{14} - B_{27} - B_{56}, \qquad E(b_7) = B_7 - B_{15} - B_{23} - B_{46}.$$

Saturated resolution III designs can be constructed for any number of factors, but they may be nonorthogonal. The following rules can be used to construct $2^K//(K+1)$ resolution III designs.

Rule 1. The ith row, $1 \le i \le K$, has the ith factor (column) at its low level and all other factors (columns) at their high levels.

Rule 2. The $(K+1)$th row has all factors (columns) at their low levels.

Rule 3. The first-order model matrix is formed by adding an initial column of positive ones. The moment matrix and its inverse are

$$(\mathbf{X'X}) = 4\,\mathbf{I}_{K+1} + (K-3)\,\mathbf{J}_{K+1},$$

$$(\mathbf{X'X})^{-1} = \frac{1}{4}\,\mathbf{I}_{K+1} - \frac{K-3}{4(K-1)^2}\,\mathbf{J}_{K+1},$$

where $\mathbf{I}_{K+1}$ is a $(K+1) \times (K+1)$ identity matrix and $\mathbf{J}_{K+1}$ is a $(K+1) \times (K+1)$ matrix with all elements equal to one.

From rule 3 the variance of any coefficient estimator is

$$\mathrm{Var}(b_i) = \left[\frac{K^2 - 3K + 4}{4(K-1)^2}\right]\sigma^2.$$

As the number of factors increases, this variance approaches the value of $\sigma^2/4$. This limiting variance is not as good as those obtained from factorial, fractional factorial, or Plackett and Burman designs. Variances of coefficient estimators for these designs approach zero as the number of factors becomes large. Estimated coefficients from nonorthogonal saturated designs are correlated because the inverse matrix is not diagonal.

Example. The main effects of five factors on a response can be investigated by a saturated $2^5//6$ design.

$$
\mathbf{X} = \begin{bmatrix}
+ & - & + & + & + & + \\
+ & + & - & + & + & + \\
+ & + & + & - & + & + \\
+ & + & + & + & - & + \\
+ & + & + & + & + & - \\
+ & - & - & - & - & -
\end{bmatrix},
$$

$$
\mathbf{Y} = \begin{bmatrix}
12.4 \\
7.6 \\
24.9 \\
2.6 \\
27.6 \\
1.4
\end{bmatrix}, \quad
\mathbf{X'Y} = \begin{bmatrix}
76.5 \\
48.9 \\
58.5 \\
23.9 \\
68.5 \\
18.5
\end{bmatrix},
$$

$$
(\mathbf{X'X})^{-1} = \frac{1}{4}\mathbf{I} - \frac{5-3}{4(5-1)^2}\mathbf{J} = \frac{8}{32}\mathbf{I} - \frac{1}{32}\mathbf{J}
$$

$$
= \frac{1}{32}\begin{bmatrix}
8 & 0 & 0 & 0 & 0 & 0 \\
0 & 8 & 0 & 0 & 0 & 0 \\
0 & 0 & 8 & 0 & 0 & 0 \\
0 & 0 & 0 & 8 & 0 & 0 \\
0 & 0 & 0 & 0 & 8 & 0 \\
0 & 0 & 0 & 0 & 0 & 8
\end{bmatrix} - \frac{1}{32}\begin{bmatrix}
1 & 1 & 1 & 1 & 1 & 1 \\
1 & 1 & 1 & 1 & 1 & 1 \\
1 & 1 & 1 & 1 & 1 & 1 \\
1 & 1 & 1 & 1 & 1 & 1 \\
1 & 1 & 1 & 1 & 1 & 1 \\
1 & 1 & 1 & 1 & 1 & 1
\end{bmatrix}
$$

$$
= \frac{1}{32}\begin{bmatrix}
7 & -1 & -1 & -1 & -1 & -1 \\
-1 & 7 & -1 & -1 & -1 & -1 \\
-1 & -1 & 7 & -1 & -1 & -1 \\
-1 & -1 & -1 & 7 & -1 & -1 \\
-1 & -1 & -1 & -1 & 7 & -1 \\
-1 & -1 & -1 & -1 & -1 & 7
\end{bmatrix},
$$

$$
\mathbf{b} = (\mathbf{X'X})^{-1}\mathbf{X'Y} = \begin{bmatrix}
9.9 \\
3.0 \\
5.4 \\
-3.2 \\
7.9 \\
-4.6
\end{bmatrix}
$$

where $\mathbf{b}' = (b_0, b_1, b_2, b_3, b_4, b_5)$. The variances and covariances of the coefficient estimators are

$$\text{Var}(b_0) = \text{Var}(b_1) = \cdots = \text{Var}(b_5) = \frac{7\sigma^2}{32},$$

$$\text{Cov}(b_i, b_j) = \frac{-\sigma^2}{32} \quad \text{for } i \neq j = 0, 1, \ldots, 5.$$

To estimate variances and covariances, we must estimate the variance σ^2 of the experimental error. This estimate may be made from past experiments or by replicating experimental units.

Fractional factorial designs, especially saturated designs, provide much information for a minimum number of experimental runs. Because the number of runs is kept to a minimum, each one is very important, and, therefore, the experimental plan must be conducted with extreme care. These designs are sensitive to missing observations, use of wrong factor levels, changes in factor levels, and oversights such as misrecording data or incorrect calculations of the response value. Running a small experiment does not imply that it should be done quickly. All the proper experimental methods such as randomization, uniform experimental units, and so on, must be followed.

Fractional factorial designs cannot satisfy all experimental situations. Situations suited to these designs occur (1) when interactions are negligible, (2) when screening many factors, and (3) when initial experiments are needed for sequential studies.

When interactions are known to be negligible, even before an experiment is run, only first-order effects need to be estimated. This knowledge may result from previous experiments or consideration of chemical and physical properties of the factors involved. Usually, assuming complete nonexistence of interactions is too strong, so interactions are assumed to be negligible. This assumption is checked with subsequent investigations.

Screening experiments are used to investigate many factors when only a few factors are expected to be significant. During the development stage of a process, many factors may be suggested that could have an effect on the response. Screening experiments can be used to select the most important factors for further study.

Many experimental situations require conducting several experiments to be run sequentially. Here, the design of an experiment may depend on results from previous experiments. Fractional factorial designs are frequently used as initial experiments to study first-order effects. Follow-up experiments are used to verify negligible interaction or model assumptions. Optimization experiments usually use a sequence of experiments to estimate changes in the response for a unit change in each factor. This rate of change is equal to estimating coefficients for the first-order effects. After using these first-order designs to get near-optimum conditions, a larger second-order design is run to characterize the response surface. Fractional factorial designs are used as first-order designs for both simplex and steepest ascent optimizations.

4.5   Half-Normal Plots

Three methods are commonly used to estimate the experimental error variance for factorial or fractional factorial designs. One method uses past experimental data. This method is rarely used because past experiments were most likely conducted under different conditions. The second method is to run replicate experimental units either by replicating the complete design or by replicating only experimental units at center points (ie, the scaled value = 0). Replicating center points requires fewer experimental units to provide a sufficient estimate of the experimental error variance. The third method concerns the case when no replicates are available. For this case, higher-order interactions are assumed to be negligible, and the experimental error variance is estimated by using the sum of the squared residuals. From the estimated variance for the experimental error, the significance of model parameters can be evaluated.

An additional graphical method, called *half-normal plots*, can also be used to examine significant effects. Half-normal plots were developed by Cuthbert Daniel (1959), and updated by Douglas A. Zahn (1975a,b), to determine significant effects in screening experiments. For these experiments, all but a few effects are expected to be negligible. Half-normal plots are presented here for screening experiments using two-level unreplicated factorial or fractional factorial experiments with 16, 32, 64, and 128 treatment combinations. From these designs the number of possible coefficients that can be estimated in the approximating model is equal to the number of treatment combinations. However, the intercept estimator, b_0, is usually significantly different from zero, so it will be excluded. This leaves $M = 15, 31, 63$, or 127 estimated coefficients to examine for significant effects. That is, $M = 2^K - 1$ for factorial designs, $M = 2^{K-1} - 1$ for half-fractional factorial designs, and so on. For convenience, coefficient estimators will be denoted here by b_i, $i = 1, 2, \cdots, M$.

If experimental errors are assumed to be i.i.d as a normal distribution with expected value zero and constant variance, then the b_i's are also normally distributed. In addition, each coefficient has the same variance for two-level orthogonal factorial and fractional factorial designs. Under the

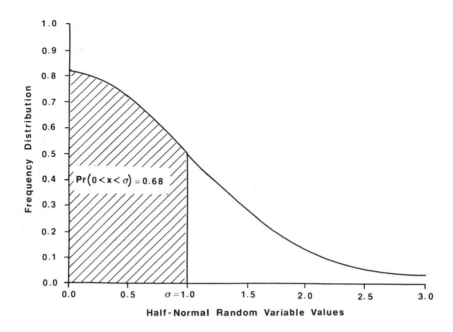

Figure 4.1.   Probability frequency distribution of a standardized half-normal random variable ($\sigma = 1.0$).

null hypothesis that all coefficients are zero, each coefficient estimator has the distribution

$$b_i \sim N(0, \sigma_b^2).$$

The variance σ_b^2 represents $Var(b_i)$ here, not the experimental error variance.

Because the coefficients' magnitudes are the important quantities, only their absolute values will be studied:

$$W_i = |b_i|, \quad i = 1, 2, \cdots, M.$$

The distribution of the W_i random variables is a half-normal distribution. The standardized half-normal distribution is illustrated in Figure 4.1. This distribution is a folded-over normal distribution. Therefore, the probability of the interval $0 < W_i < \sigma_b$ is twice the probability of $0 < b_i < \sigma_b$:

$$Pr\,(\,0 < b_i < \sigma_b) = 0.34, \quad Pr\,(\,0 < W_i < \sigma_b) = 0.68.$$

The 0.68 percentile for the half-normal distribution is σ_b. Order the absolute values in ascending order $W_1 < W_2 < \cdots < W_M$. Ordered values will be denoted by subscript j rather than i (ie, W_j). The ordered statistic W_r that is closest to the 0.68 percentile is an estimate of σ_b. The values of r for different values of M are

$$M: \quad 15 \ 31 \ 63 \ 127$$
$$r: \quad 11 \ 22 \ 44 \ \ 87$$

The inherent assumption here is that coefficients associated with the first r-values of W_j are not significantly different from zero. Half-normal plots are useful for detecting a maximum of M-r real effects in screening experiments. For example, four or fewer real effects can be detected when $M = 15$, and nine or fewer real effects can be detected when $M = 31$.

The expected value of W_j under the null hypothesis that all coefficients are nonsignificant is

$$E(W_j) = \sigma_b \, H_{jM},$$

where H_{jM} is the expected value of the jth ordered statistic in a random sample of size M from the standardized half-normal distribution (Zahn 1975a). If the W_j's are plotted versus the H_{jM}'s, the plot should be a straight line passing through the origin. A W_j, $j > r$, would not conform to a line if its associated coefficient were significantly different from zero. Then the value of W_j would be greater than a multiple of estimated standard deviations, W_r's,

$$W_j > C_j W_r.$$

Multiplying factors C_j have been calculated by Zahn (1975a) and are reproduced in Table 4.2 for error rates of 0.05 and 0.20. Error rates are probabilities of falsely declaring at least one coefficient to be significant when simultaneously testing M-r coefficients.

Values of $C_j W_r$ can be plotted on the half-normal plot and connected to form a line called a "guardrail." Those values of W_j, $j > r$, above the guardrail are declared significant. From the remaining nonsignificant values, the final standard deviation estimator S_F is determined. Let L represent the number of nonsignificant estimates. If the L values of W_j are plotted, then they should fall on a line through the origin with a slope of σ_b, because their expected values are $E(W_j) = \sigma_b H_{jL}$. The estimator S_F is the least squares regression slope through the origin of the smallest $m = 0.7(L+1)$ values of the L nonsignificant estimates. The value m was empirically determined to be best for the estimator S_F.

To apply half-normal plotting, the H_{jM}'s are approximated by Z_j values that correspond to $0.5 + (j-0.5)/2M$ percentile points of the standard

TABLE 4.2 Multiplying factors, C_j's, for error rates 0.05 and 0.20 used to compute guardrails (kind permission of the author, Douglas A. Zahn. 1975. *Technometrics* 17:189–200)

	N = 15 Error rates			N = 31 Error rates	
j	0.05	0.20	j	0.05	0.20
12	2.065	1.533	27	2.615	2.133
13	2.427	1.866	28	2.807	2.288
14	2.840	2.177	29	2.992	2.439
15	3.230	2.470	30	3.173	2.586
			31	3.351	2.730

	N = 63 Error rates			N = 127 Error rates	
j	0.05	0.20	j	0.05	0.20
58	3.030	2.570	121	3.396	2.960
59	3.120	2.647	122	3.442	2.999
60	3.209	2.722	123	3.487	3.039
61	3.297	2.797	124	3.532	3.078
62	3.384	2.872	125	3.576	3.116
63	3.470	2.945	126	3.620	3.155
			127	3.663	3.193

normal distribution [ie, $\Pr (Z < Z_j) = 0.5 + (j-0.5)/2M$]. Half normal plots can be made by using the following rules.

Rule 1. Order the absolute values of the estimated coefficients, $W_j = |b_j|$, such that $W_1 < W_2 < \cdots < W_M$.

Rule 2. Plot W_j versus Z_j, where Z_j, is the value corresponding to the $0.5 + (j-0.5)/2M$ percentile point of the standard normal distribution.

Rule 3. Make an initial estimate of the standard deviation by the rth-order statistic, W_r, where $r = 11, 22, 44,$ and 87 for $M = 15, 31, 63,$ and 127, respectively.

Rule 4. Plot the guardrail values $W_r C_j, j = r+1, \cdots, M$, for a selected error rate. Declare those estimated coefficients corresponding to the W_j's above the guardrails to be significantly different from zero.

Rule 5. Replot the L nonsignificant estimates versus the Z_j's for the $0.5 + (j-0.5)/2L$ percentile points of the standard normal distribution.

Rule 6. Compute S_F. The final standard deviation estimate of the coefficient estimators is computed from a subset of the L nonsignificant estimates. Compute S_F from the smallest $m = 0.7(L+1)$ values of the nonsignificant estimates. The estimator S_F is the slope of the least square line through the origin and is calculated by

$$S_F = \frac{\sum_{j=1}^{m} W_j Z_j}{\sum_{j=1}^{m} Z_j^2} \, .$$

Half-normal plotting can be illustrated with data from W. Lindberg, E. Johansson, and K. Johansson (1981). These authors were trying to find a method in routine work that would accomplish a separation good enough for quantification of four alkaloids: morphine, codeine, noscapine, and papaverine in a commercial solution. In this study, the effects of four factors were investigated with a full 2^4 factorial on capacity ratio measurements for reversed-phase ion-pair chromatography. The four factors are given in the following table:

Factor		Levels	
A:	Methanol–water ratio (v:v)	32:68	38:62
B:	pH	2.0	4.0
C:	Buffer concentration (M)	0.01	0.09
D:	Camphorsulphonic acid (M)	0.000	0.010

Estimated main effects and interactions, which may be called contrasts, are given in Table 4.3. A BASIC computer program is given in Table A.3 (see Appendix: BASIC Programs) to calculate abscissa values, guardrail values, and the final estimated standard deviation of the contrast estimators. Program output for papaverine is given in Table 4.4, and the corresponding half-normal plot is illustrated in Figure 4.2.

The half-normal plot shows that contrasts A, D, and AD are significant with a 20% guardrail. Analysis of noscapine contrasts would show similar results. Contrasts A and D would be significant for codeine, and contrast D would be significant for morphine. The 20% guardrail may indicate too many significant contrasts. However, because screening experiments are used to pick factors for further studies, being slightly conservative about declaring nonsignificant contrasts may be a better policy than using a 5% guardrail. The authors of this study concluded that contrasts A, D, and AD were the most important and proceeded to examine the response surfaces for the four alkaloids.

TABLE 4.3. Contrast values (main effects and interactions) estimated from a full two-level factorial design investigating effects of four factors on capacity ratios (Lindberg, Johansson, and Johansson 1981)

Contrast	Papaverine	Noscapine	Codeine	Morphine
A	−4.893	−3.305	−0.238	−0.093
B	−0.293	−0.163	−0.039	−0.020
C	−0.779	−0.696	−0.086	−0.046
D	3.974	3.155	0.398	0.219
AB	0.351	0.300	0.016	0.006
AC	0.283	0.206	0.014	0.004
AD	−2.058	−1.460	−0.124	−0.052
BC	−0.030	−0.016	−0.009	−0.006
BD	−0.925	−0.770	−0.093	−0.052
CD	−0.869	−0.654	−0.104	−0.063
ABC	−0.041	0.019	0.003	0.003
ABD	0.556	0.393	0.032	0.014
ACD	0.295	0.269	0.023	0.013
BCD	−0.135	−0.166	−0.021	−0.014
ABCD	0.001	−0.026	0.004	0.004

TABLE 4.4. Output of the half-normal BASIC computer program for the papaverine contrasts

Half-normal plot values		
Order	Contrast	Z-Value
1	0.001	0.042
2	0.03	0.125
3	0.041	0.21
4	0.135	0.296
5	0.283	0.385
6	0.293	0.477
7	0.295	0.573
8	0.351	0.674
9	0.556	0.783
10	0.779	0.903
11	0.869	1.036
12	0.925	1.192
13	2.058	1.383
14	3.974	1.645
15	4.893	2.128
Guardrail values for the 5% and 20% significance levels		
Num	5% Level	20% Level
12	1.794485	1.332177
13	2.109063	1.621554
14	2.46796	1.891813
15	2.80687	2.14643

Number of nonsignificant contrasts at 20% level = 12.
Estimated standard deviation of coefficients = 0.4549598.

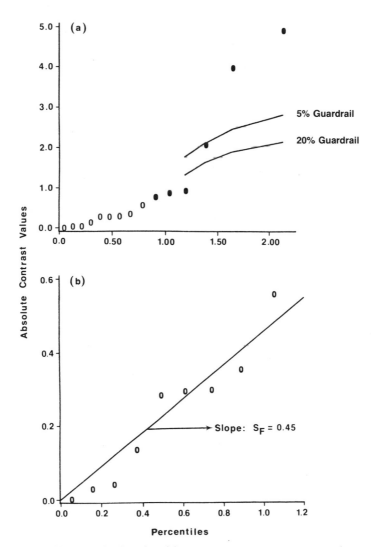

Figure 4.2.   (a) Half-normal plot for fifteen papaverine contrast values. (b) Final contrast standard deviation estimated by the slope of a line through the origin fitted to a subset (open circles) of the non-significant contrasts.

5

Optimization Experiments

Most laboratory procedures, at some time during their development, have been subjected to optimization. More than likely this optimization procedure has been the old one-factor-at-a-time method, because that is the way generations of experimental scientists have been taught. This method is still popular, although the study of experimental design has been systematized into a branch of statistics since 1925 when R. A. Fisher did his original work (Fisher 1925). Chapters on experimental design can be found in most standard texts on statistics, and specific topics have been discussed in many articles in the various journals of statistics and mathematics.

To apply statistical optimization methods to practical laboratory problems, one must first have a concept of the meaning of optimization. Optimization is generally taken to mean finding the best possible method of carrying out some operation. In practice, considerations are made for such things as time, funding, availability of material and personnel. Therefore, an optimization study should result in a practical solution that may fall somewhat short of maximizing any one specific goal, but still do the best job within a particular set of constraints.

Optimization experiments are a special class of statistically designed experiments that provide a means of reaching optimum operating conditions by allowing the experimenter to change more than one factor at a time. Multifactor experiments can usually find the optimum with a smaller number of experimental runs than would be required by the unifactor approach. Moreover, information can also be obtained about the effects of interactions among the factors on the response that is not available with one-factor-at-a-time experiments.

Optimization is formulated as a maximization of a response, Y. This formulation is not restrictive, because if the minimum is required, a new response can be defined by $Z = -Y$. The maximum of all possible response values is termed a *global maximum*. Usually the maximum value found cannot be determined to be a global maximum. Most likely, the value will be a maximum in a region specified by the factor ranges and may not be unique. Two sequential methods to find maximum values, *steepest ascent* and *simplex optimization*, are described in this chapter. Steepest ascent uses the result of an initial experiment to calculate the gradient or derivatives of a response to determine the direction of movement. Derivatives are not required for the simplex method. Responses from an initial experiment need only be ranked from worst to best to determine the next simplex experiment.

5.1 Empirical Optimization Conditions

Five steps should be considered in planning an optimization experiment:
1. Select the response(s) to be maximized or minimized.
2. Select factors to be varied in the experiment.
3. Select factor levels and step size.
4. Select the initial experiment.
5. Select the method to determine subsequent experiments.

5.1.1 Response Selection

Empirical optimizations are methods for optimizing a response measurement in the presence of experimental error as opposed to optimizing a mathematical function. These methods are usually only practical for responses that can be made in a short period of time. However, there are a plethora of instruments that give quick and reliable responses. Empirical optimization methods envision a chemist in the laboratory running an experiment, reading the results, and then running the next experiment. The complete sequence of optimization experiments should be finished in a day or in a week. An experimental run that takes days or weeks to complete is not suitable for sequential experiments and requires a different strategy.

Recall that an observed response is the sum of a true value plus experimental error. The rate of advancement towards the optimum will depend both on the form of the response function and the standard deviation of the error. For example, simplex optimization advances at a rate that is inversely proportional to the standard deviation of the error (Spendley,

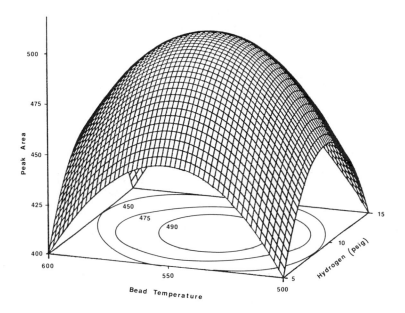

Figure 5.1. Peak area response surface and contour plots.

Hext, and Himsworth 1962). Therefore, an optimization program based on a response with a small experimental error will be more efficient (ie, find an optimum more quickly) than one based on a response with a larger error.

True response values are assumed to be a function of factor levels, $F(x_1, \ldots, x_K)$, which represent a continuous surface. It is convenient to visualize the response surface function geometrically by a response surface plot. Suppose peak area for nitrogen compounds using a rubidium bead type of nitrogen–phosphorus gas chromatographic detector is related to bead temperature and hydrogen pressure by the response surface in Figure 5.1. This plot gives a picture of the behavior of peak areas for different levels of bead temperature and hydrogen pressure. An alternative method of representing response surfaces is by *contour plots*, also shown in Figure 5.1. Contour plots are lines of equal response for different levels of the factors. These plots are similar to contour maps used to represent mountain altitudes or to represent isobars on a weather map. To graph a two-factor response surface requires three axes, but use of contour plots reduces a response surface plot by one dimension. Therefore, a two-factor response surface may be represented by a two-dimensional contour plot.

If the additional factor of air pressure also affected peak area, several contour plots, such as Figure 5.2, might represent the response surface. Here a contour plot is drawn for each level of the third factor, giving an

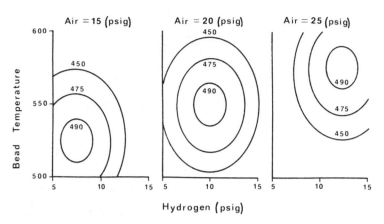

Figure 5.2.   Peak area contours for three factors.

impression of the effect of all three factors. The object of optimization is to reach the response surface summit or locate factor levels that enclose the response in a high contour level.

Frequently, chemists want to maximize several measurements while minimizing others. Section 2.6 indicates two methods for dealing with multiple responses: superimposing contour diagrams or constructing desirability coefficients. Both methods have been successfully used in practice.

5.1.2   Factor Selection

Factor selection means deciding which factors have the largest effect on the responses and will be varied during the experiment. However, factors such as sample volume, reagent volume, and reagent concentration must be used with care because optimization of unnormalized responses may proceed toward higher factor levels due only to increases in quantity. For example, if absorbance is used as a measure of sensitivity, either it should be divided by the sample volume or the sample volume should be fixed.

Factors that cannot be varied are assigned a fixed value, and optimization is done with the restriction that these factors are at an assigned value. Other variables that cannot be varied or assigned a fixed value, such as the time of day, may be used as blocking variables.

5.1.3   Factor Levels and Step Size

Factor levels determine the step size of sequential optimization experiments and, therefore, influence the rate of finding the optimum. Response changes are not an absolute measure of factor effects but depend on the step size selected for the factors. Step size should be large enough to allow effects of factors to show up but small enough so that factor effects are approximately linear over each step. In practice, factor levels and step size are determined by good judgment. A step size for each factor is usually chosen to cause a comparable change in the response. Bad judgment can be repaired by pausing to "take stock" after some initial experiments and then updating step size or using variable step size, which can be done in simplex optimization.

Suppose an initial 2^2 factorial experiment will be run for factors concentration (g/L) and time (min) with a 0.5 g/L increase in concentration comparable to a 10-min increase in time. The response depends on factors X_1 and X_2, and the initial design is given in Table 5.1.

TABLE 5.1. Example of an initial two-level factorial design for two factors

Concentration (g/L)	Time (min)
14.5	50
14.5	70
15.5	50
15.5	70

Coding the factors to high and low values,

$$X_1 = \frac{g/L - 15}{0.5}, \qquad X_2 = \frac{min - 60}{10},$$

defines the design space. A change of one unit in the design space is called the *step size* and corresponds to 0.5 g/L and 10 min for the two factors. Step size is one-half the range for the low and high levels. The zero level in the design space is the average of the low and high levels and is called the *base level*, corresponding to 15 g/L and 60 min for this example. The base level is chosen at standard operating conditions or at conditions felt to be optimum. Thus, optimization moves from base level conditions to conditions that will improve the response, and the rate of optimization depends on the chosen step size.

5.1.4 Initial Experiment

Initial experiments for the steepest ascent method have two purposes:
1. Fit a first-order equation to approximate the response near the base level.

2. Provide a test for the adequacy of the approximating model within limits of the experimental error.

The first objective can be accomplished by using 2^K factorial and fractional factorial designs or first-order designs, such as Plackett and Burman designs that were discussed in Chapter 4.

The second objective can be met by replicating design points or adding center points to the initial experiment. The sum of the squared residuals (residuals = observed − predicted) can then be partitioned into a sum of squares due to lack of fit and a sum of squares due to pure error, which is calculated from replicated center points (base level points) or from replicated initial design points. These two sums of squares are compared with an F-test (calculation details will be given in the subsection "Steepest Ascent Example") to judge if a linear model is adequate. If no significant lack of fit is found, sequential experiments are performed along the path of steepest ascent.

A significant lack of fit may indicate (1) significant curvature is present, or (2) pure error is underestimated. Significant curvature means the step size is too large or the approximating model needs higher-order terms. In this situation, either do another initial experiment using a smaller step size or augment the initial experiment with additional runs to explore a higher-order surface. Another reason for significant lack of fit to occur is that the replicated points were not properly randomized. For example, if all the replicated center points were measured simultaneously for convenience, pure error may be underestimated due to correlation among these observations.

Initial construction of a simplex design is based on regular geometrical simplex figures. Regular simplex designs are triangles in two dimensions, tetrahedrons in three dimensions, and figures with $K+1$ vertices in K dimensions. Brooks and Mickey (1961) showed that simplex designs, without replication, are optimal for estimation of first-order effects in the presence of error. However, the added attraction of simplex designs is that a new simplex design can be formed by discarding a vertex and adding another vertex that is a reflection through the face opposite to the discarded one. A simplex optimization to maximize a chemical response depending on K factors starts with an initial simplex design of $K+1$ combinations of the factors. Orientation of simplex designs in the factor space does not affect their efficiency, but it is convenient to orient one edge parallel to a factor axis. Table 5.2 (Long 1969) is used to construct simplex designs for up to ten factors. This table specifies fractions of step sizes to be taken from the base level. A triangular simplex for two factors, A and B, will require vertices 1, 2, and 3. A three-factor simplex design is a tetrahedron, so vertices 1, 2, 3, and 4 are required for factors A, B, and C.

TABLE 5.2. The step-size fraction added to the base level for initial simplex designs (by kind permission of D. E. Long, 1969. *Anal. Chem. Acta* 46:193–206)

Vertex no.	Factors									
	A	*B*	*C*	*D*	*E*	*F*	*G*	*H*	*I*	*J*
1	0.00	0.00	0.00	0.00	0.00	0.00	0.00	0.00	0.00	0.00
2	1.00	0.00	0.00	0.00	0.00	0.00	0.00	0.00	0.00	0.00
3	0.50	0.87	0.00	0.00	0.00	0.00	0.00	0.00	0.00	0.00
4	0.50	0.29	0.82	0.00	0.00	0.00	0.00	0.00	0.00	0.00
5	0.50	0.29	0.20	0.79	0.00	0.00	0.00	0.00	0.00	0.00
6	0.50	0.29	0.20	0.16	0.78	0.00	0.00	0.00	0.00	0.00
7	0.50	0.29	0.20	0.16	0.13	0.76	0.00	0.00	0.00	0.00
8	0.50	0.29	0.20	0.16	0.13	0.11	0.76	0.00	0.00	0.00
9	0.50	0.29	0.20	0.16	0.13	0.11	0.09	0.75	0.00	0.00
10	0.50	0.29	0.20	0.16	0.13	0.11	0.09	0.08	0.75	0.00
11	0.50	0.29	0.20	0.16	0.13	0.11	0.09	0.08	0.08	0.74

For example, suppose a normalized response depends on the three factors temperature, amount of solution, and pH. An initial simplex design can be constructed by using the first four vertices in Table 5.2 for factors A, B, and C. Table 5.3 shows the initial design step sizes indicated below each factor.

TABLE 5.3. Example of an initial design for a three-factor simplex design

Vertex	Temperature (10°C)	Solution (5 ml)	pH (0.5)
1	110	100	5.5
2	120	100	5.5
3	115	104.4	5.5
4	115	101.5	5.9

Because zeros in Table 5.2 indicate that no fractions are to be added to the base level, vertex 1 is the set of base level values. Vertex 2 is calculated by adding one step size to the temperature factor (110 + 10) while the other two factors remain at their base level values. Vertex 3 is calculated by adding 0.5 step-size fraction to temperature base level $(110 + 0.5 \times 10)$, adding 0.87 step-size fraction to solution base level $(100 + 0.87 \times 5)$, with no change in the pH base level. All factors are incremented for vertex 4 by adding the step-size fractions 0.5, 0.29, and 0.82 to the base levels for temperature, solution, and pH, respectively.

5.1.5   Optimization Method

After an initial experiment, a method is needed to move from the base level to another point in the factor space having a larger response. Two methods are proposed that can be easily applied in the laboratory: steepest ascent and simplex optimization.

Steepest ascent fits a first-order polynomial estimated from the initial experiment to a subregion of the factor space. The estimated coefficients show the relative amounts by which the factors must be varied to give maximum increase in the response. They also determine the direction, at right angles to tangents to the true response contours, for performing subsequent experiments. Subsequent experiments are continued until little progress is made or there is evidence that curvature must be taken into account. At this point another initial experiment is performed, and steepest ascent is repeated, or a second-order design can be used to estimate the response surface. Often, factor values that give the largest observed response in a steepest ascent experiment may be adequate for meeting the experimental objectives.

For simplex optimization the location of subsequent experiments is determined by discarding the vertex at the lowest response and replacing it with its mirror image through that face of the simplex that is opposite the discarded vertex. Simplex optimization has several attractive properties: (1) calculations involved are trivial; (2) direction of advance depends solely on ranking the responses; (3) simplex designs for K factors require only $K + 1$ design points; and (4) to move to a new factor space requires only one experimental run. Given a fixed optimum, the system of simplexes will eventually circle near it. The closeness to the optimum will depend on the step size.

The most popular choice of the two methods has been simplex optimization. Deming and Morgan (1983) list 189 applications of simplex optimization, which include several modifications to the basic method. Simplex optimization is easily applied to a variety of laboratory situations. However, simplex optimization is not usually as efficient as steepest ascent because the simplex method uses only ranks, not magnitudes, of responses.

Often, steepest ascent shows superior performance for response surfaces that have continuous derivatives (Spendley et al. 1962).

Chemists should keep in mind that both methods are linear techniques. Continued application of linear techniques with progressively reduced step size is inherently self-defeating. The objective is to get near the optimum or a region of stationary values with larger responses. If the results of these linear techniques are unsatisfactory, further progress may be made by doing experiments to fit higher-order response surfaces.

5.2   Steepest Ascent

Box and Wilson (1951) showed how to apply steepest ascent to optimize responses in the presence of experimental error. In general, the problem involves moving from point O (base level) to a point P in the factor space, so that the gain in response is a maximum. If the surface containing the points O and P can be approximated by a plane, the coordinates at P of maximum gain are proportional to their corresponding first-order partial derivatives. Therefore, if a response, Y, is related to values of K factors by the function

$$Y = F(x_1, x_2, \ldots, x_K) + \text{error},$$

and this function can be approximated in a subregion of the factor space by a first-order approximation

$$F(x_1, x_2, \ldots, x_K) \approx B_0 + B_1 x_1 + B_2 x_2 + \cdots + B_K x_K,$$

then at P,

$$x_j \propto \frac{\partial F}{\partial x_j} = B_j.$$

By incrementing the factor values at O proportional to their slopes, the chemist performs sequential experiments in a direction at right angles to tangents of the true response contours, which is the direction of steepest ascent.

5.2.1   Steepest Ascent Step Size

To move to a new experimental point, we must update the jth factor value by a Δx_j (ie, $x_j + \Delta x_j$). This update is proportional to the corresponding slope by the constant $1/2R$:

$$\Delta x_j = \frac{B_j}{2R}.$$

The proportional constant R is chosen by selecting a convenient increment to change one factor, say the kth factor:

$$R = \frac{B_k}{2 \Delta x_k}.$$

The units for slope and Δx_k must be compatible. If the slope B_k is estimated by using factor levels in coded units, then Δx_k must also be in coded units. For example, a 2^2 experiment was used to estimate the slopes for temperature and concentration with factor levels of

Temperature (°C):	80(−1)	90(+1)
Concentration (mg/g) :	40(−1)	60(+1)

Suppose, for a yield response, the estimated slopes derived by using coded factor levels are $b_1 = -2.5$ and $b_2 = 1.5$. (Note, for yield to increase, temperature must decrease because b_1 is negative.) To move along the path of steepest ascent, we change, **for convenience**, the temperature by $\Delta x_1 = -2°C$, which is $\Delta x_1 = (-2°C)/(5°C) = -0.4$ in coded units. The proportional constant has the value

$$R = \frac{-2.5}{2(-0.4)} = 3.125.$$

Therefore, concentration with 10 mg/g equal to one coded **unit should be** incremented by

$$\Delta x_2 = \frac{10 \, b_2}{2R} = \frac{(10)\,(1.5)}{(6.25)} = 2.4 \text{ mg/g.}$$

Moving along the path of steepest ascent is done by decreasing the temperature by 2°C and increasing concentration by 2.4 mg/g.

5.2.2   Stopping Rules

Before beginning steepest ascent, the chemist must consider conditions for stopping and restarting the process. Three general rules for stopping steepest ascent are the following
1. Responses decrease for two consecutive experiments.
2. Responses do not change for two consecutive experiments.

3. The difference between observed and predicted responses is too large.

Rule 1.   This rule is to prevent steepest ascent from going too far. Steepest ascent may reach a maximum and start decreasing. The first rule can be easily applied by examining the magnitude of the responses for consecutive experiments.

Rule 2.   Lack of change in response indicates that steepest ascent is on a plateau, and little improvement may be expected by continuing. At this point, a second-order design may be appropriate, or it may be appropriate to stop if the response is considered satisfactory. Let Y_j and Y_{j+1} represent two consecutive responses in a steepest ascent experiment. The responses are assumed to be independent and to have constant variance, $\text{Var}(Y)$; therefore, the variance of their difference is

$$\text{Var}(Y_{j+1} - Y_j) = \text{Var}(Y_{j+1}) + \text{Var}(Y_j) = 2\text{Var}(Y).$$

No change in the response for consecutive experiments can mean that the response difference in consecutive experiments is less than a multiple of the standard deviation of the difference. One choice of this multiplier is based on the observation that, for a standardized normal distribution, $\text{Pr}(-0.68 < Z < 0.68) = 0.5$. Hence, no response change is judged by the criterion

$$|Y_{j+1} - Y_j| < 0.68 \sqrt{2\text{Var}(Y)}.$$

To use this criterion, we need an estimate of the response variance. This variance estimate is usually made by replicated experimental units in the initial experiment.

Rule 3.   This rule is applied to detect curvature of the response surface. A large difference between observed and predicted responses along the path of steepest ascent would indicate that the response surface can no longer be approximated by a plane. A new steepest ascent path to change direction or a second-order design would be required. A large difference between an observed response, Y, and a predicted response, Y_P, along the steepest ascent path will also be judged by a multiple of the standard deviation of their difference. The variance of their difference is

$$\text{Var}(Y - Y_P) = \text{Var}(Y) + \text{Var}(Y_P).$$

Calculation of $\text{Var}(Y_P)$ is a problem because the variance of a predicted value depends on the values of the K factors.

To estimate Y_P, assume that a factorial or a fractional factorial experiment augmented by replicated center points is used. The factorial portions of the experiment are used to estimate the linear coefficients in the predicted model (ie, Y_P). Replicated center points are used to estimate $Var(Y)$ and to check for lack of fit. Let the number of experimental points be denoted by

F = number of factorial or fractional factorial points,
C = number of center points.

For this experiment, factor levels are coded (ie, -1 = low level and $+1$ = high level), and experimental points for predicted responses are defined by this scaling. Then, using the variance formula in Chapter 4, we obtain the variance of the predicted response at a point with factor values ($\pm a_1, \pm a_2, \ldots, \pm a_K$)

$$Var(Y_p) = \left(\frac{1}{F+C} + \sum_{j=1}^{K} \frac{a_j^2}{F} \right) Var(Y),$$

so

$$Var(Y-Y_p) = \left(1 + \frac{1}{F+C} + \sum_{j=1}^{K} \frac{a_j^2}{F} \right) Var(Y).$$

A multiplier of the standard deviation of the difference is based on the probability of a normal random variable, $Pr(-2<Z<2) \sim 0.95$. Therefore, a criterion for too large a difference between observed and predicted response is

$$|Y - Y_p| > 2\sqrt{Var(Y-Y_p)}.$$

For example, the coefficients for a linear model of peak area are estimated by a 2^3 ($F = 8$) factorial augmented by $C = 4$ center points. The three detector factors of hydrogen pressure, air pressure, and bead current are coded as in Table 5.4.

TABLE 5.4. Factorial levels for a three-factor design

Coded levels	-1	0	$+1$	Step size
H_2 pressure	6	12	18	6
Air pressure	20	30	40	10
Bead current	575	625	675	50

An experimental point (24,15,750) in the factor space is coded as $(+2,-1.5,2.5)$, and the variance of the difference between observed and predicted response can be calculated as follows:

$$\sum_{j=1}^{3} a_j^2 = (2)^2 + (-1.5)^2 + (2.5)^2 = 12.50,$$

$$Var(Y-Y_p) = \left(1 + \frac{1}{12} + \frac{12.50}{8}\right) Var(Y) = 2.65 Var(Y).$$

If the response variance estimate from the center points is $S^2 = 4.00$, then rule 3 would stop steepest ascent if

$$|Y-Y_p| > 2\sqrt{2.65(4.00)} = 6.51.$$

During steepest ascent, new factorial designs may be used to update coefficient estimates. These new designs may use factor codings different from the original design, but to calculate rule 3 the scaling from the design used to estimate response variance should be used. For rules 2 and 3, multipliers are based on convenience and reference to standardized normal distribution. If these are considered inappropriate, other multipliers may be used, such as a t-statistic in rule 3 based on the number of degrees of freedom used to estimate $Var(Y)$.

5.2.3   Steepest Ascent Example

For this example the objective of a steepest ascent experiment is to optimize yield of a process in terms of temperature and a solution concentration of a process component. Present conditions give a product yield of about 50 mg/g at 20°C and 20 mg/g of reagent. A linear model is considered to represent yield adequately near present conditions. A 2^2 factorial with factor levels given in Table 5.5 is used to estimate the linear coefficients and is augmented by three center points to estimate response variance and examine lack of fit. The yield response and temperature and concentration factors are represented by the following notation: Y = yield (mg/g), T = temperature (°C), C = concentration (mg/g).

TABLE 5.5. Factor levels for the yield experiment

Coded units	−1	0	+1	Step size
Temperature	15	20	25	5
Concentration	17.5	20	22.5	2.5

$X_1 = (T\text{-}20)/5$ and $X_2 = (C\text{-}20)/2.5$.

A 2^2 factorial with three center points was performed in random order, and the results are listed in Table 5.6.

TABLE 5.6. Initial design for the yield experiment. A two-level factorial with three center points

Run	Order	T	C	X_1	X_2	Y
1	1	25	22.5	+1	+1	52.01
2	3	25	17.5	+1	−1	49.82
3	5	15	22.5	−1	+1	46.50
4	7	15	17.5	−1	−1	44.57
5	2	20	20	0	0	49.00
6	6	20	20	0	0	47.67
7	4	20	20	0	0	51.07
					Total	340.64
					Average	48.66

This experiment is used to estimate the predicted response

$$Y_P = b_0 + b_1 X_1 + b_2 X_2.$$

The general least squares matrix solution for linear models is used to estimate the model parameters. For factorial designs augmented by center points, this solution implies that the intercept estimate is the average of the responses, and coefficient estimates shown in Table 5.7 are the average of the difference for responses at low and high factor levels.

TABLE 5.7. Slope estimates from the initial design of the yield experiment

Response	T	C
High level	101.83	98.51
Low level	91.07	94.39
Difference	10.76	4.12
Factorial points	4	4
Slopes	2.69	1.03

The prediction model is

$$Y_p = 48.66 + 2.69X_1 + 1.03X_2.$$

To judge the adequacy of this model, we will examine the sum of squared differences between the observed and predicted values (ie, residuals) given in Table 5.8.

TABLE 5.8. Calculations for lack of fit test

Run	X_1	X_2	Y	Y_p	Residuals
1	+1	+1	52.01	52.38	−0.37
2	+1	−1	49.82	50.32	−0.50
3	−1	+1	46.50	47.00	−0.50
4	−1	−1	44.57	44.94	−0.37
5	0	0	49.00	48.66	0.34
6	0	0	47.67	48.66	−0.99
7	0	0	51.07	48.66	2.41
	Center points average = 49.25			SSR = 7.68	

The residual sum of squares (SSR) can be partitioned into sum of squares due to lack of fit (SSLOF) and sum of squares due to pure error (SSPE). Pure error sum of squares can be calculated from the replicated center points, and SSLOF is found from the relation

$$SSLOF = SSR - SSPE.$$
$$SSPE = (49.00-49.25)^2 + (47.67-49.25)^2 + (51.07-49.25)^2,$$
$$SSPE = 5.87,$$
$$SSLOF = SSR - SSPE = 7.68 - 5.87 = 1.81.$$

If the error of the response is assumed to be i.i.d. $N(0, Var(Y))$, an F-statistic can be calculated from pure error and lack of fit mean squares. Mean squares are sums of squares divided by their degrees of freedom (df = number of points − number of estimated parameters). For the residuals, the intercept and two slopes are the estimated parameters, and for the pure error the mean response at the center is the estimated parameter.

$$dfR = 7 - 3 = 4, \quad dfPE = 3 - 1 = 2,$$
$$dfLOF = dfR - dfPE = 4 - 2 = 2.$$

$$F(2,2) = \frac{SSLOF/dfLOF}{SSPE/dfPE} = \frac{1.81/2}{5.87/2,}$$

$$F(2,2) = 0.31.$$

For a significant lack of fit, the expected F-statistic would be large. This calculated F-value would have to have a value as large as 19.00 to be significant at the 5% significance level. This inference is made by comparing the calculated F-value with tabulated percentile points for the F-distribution (Snedecor and Cochran 1967) with (2,2) degrees of freedom. Because there is no indication of lack of fit by a plane to this subregion, steepest ascent may be used to seek the optimum. In addition, the mean square for pure error can be used as an estimate of Var(Y) or

$$S^2 = \hat{Var}(Y) = 2.94, \quad S = 1.71.$$

This estimate is used to calculate stopping rules 2 and 3.

Before doing steepest ascent experiments, the step size for concentration is calculated in Table 5.9 to be comparable to a desired change of 10°C in temperature. Recall that slopes were estimated from the coded variables.

TABLE 5.9. First step-size calculations for steepest ascent for the yield experiment

Calculations	T	C
One coded unit	5	2.5
Desired change	10	
Change in coded units	2	
Slopes	2.69	1.03
Constant $R = 2.69/2(2)$	0.6725	
Step size	10	1.91

For experiments with the objective of finding the minimum, the step sizes would have the opposite signs of their slopes.

Sequential experiments along the path of steepest ascent are shown in Table 5.10.

TABLE 5.10. First steepest ascent path for the yield experiment

Run	T	C	X_1	X_2	Y	ΔY (1.65)*	Y_p	D (Y-Y$_p$)	$2S_D$
Base	20	20							
Step	10	1.9							
8	30	21.9	2	0.76	55.65	6.40	54.82	0.83	5.07
9	40	23.8	4	1.52	58.97	3.32	60.99	− 2.01	8.12
10	50	25.7	6	2.28	62.71	3.74	67.15	− 4.44	11.54
11	60	27.6	8	3.04	62.11	− 0.60	73.31	− 11.20	15.08
12	70	29.5	10	3.80	57.14	− 4.97	79.47	− 22.33	18.67

*$0.68 \sqrt{2Var(Y)} = 1.65$.

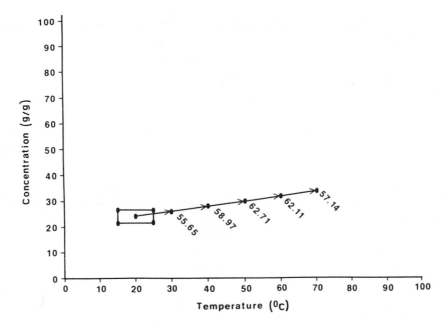

Figure 5.3.   Initial steepest ascent path for yield experiment.

Figure 5.3 shows the initial experiment and path of steepest ascent. The sequential experiments were stopped on run 12 because responses decreased on runs 11 and 12 (rule 2), and $| Y - Y_p |$ is more than twice its standard deviation (rule 3).

A new 2^2 factorial was run at the base level of (50°C, 25 mg/g) with new coded values given in Table 5.11.

TABLE 5.11.   Factor levels for the second two-level factorial design for the yield experiment

Coded units	-1	0	$+1$	Step size
Temperature (°C)	40	50	60	10
Concentration (mg/g)	20	25	30	5

The coded factors for this factorial design are $X_1 = (T-50)/10$ and $X_2 = (C-25)/5$, and the results are given in Table 5.12.

The slopes are estimated in Table 5.13.

Therefore, the new prediction model is

$$Y_P = 58.69 - 1.51X_1 + 4.99X_2.$$

Recalculating the step size in Table 5.14 gives a new steepest ascent path.

TABLE 5.12. Results from the second two-level factorial design for the yield experiment

Run	T	C	X_1	X_2	Y
13	60	30	+1	+1	63.81
14	60	20	+1	-1	50.56
15	40	30	-1	+1	63.55
16	40	20	-1	-1	56.85
				Average	= 58.69

TABLE 5.13. Slope estimates from the second factorial design for the yield experiment

Response	T	C
High level	114.37	127.36
Low level	120.40	107.41
Difference	-6.03	19.95
No. of points	4	4
Slopes	-1.51	4.99

TABLE 5.14. Second step-size calculations for steepest asccent for the yield experiment

Calculations	T	C
One coded unit	10	5
Desired change	-5	
Change in coded units	-0.5	
Slopes	-1.51	4.99
Constant R = -1.51/2(-0.5)	-1.51	
Step size	-5	1.7

A second set of sequential experiments can now be performed along this new path given in Table 5.15.

Figure 5.4 shows that this series of steepest ascent experiments moved back in the direction of the initial design. This move indicates that steepest ascent may correct inaccurate slope estimates in later experiments. This experimental series was terminated because the responses decreased in runs 19 and 20 (rule 1).

Another 2^2 factorial was run centered at (35° C, 30 mg/g) to reestimate

TABLE 5.15. Second steepest ascent path for the yield experiment

Run	T	C	X_1	X_2	Y	ΔY (1.65)	Y_p	D $(Y-Y_p)$	$2S_D$
Base	50	25							
Step	−5	1.7							
17	45	26.7	−0.5	0.34	62.09		61.14	0.95	10.33
18	40	28.4	−1.0	0.68	63.99	1.90	63.59	0.40	9.61
19	35	30.1	−1.5	1.02	61.87	−2.12	66.04	−4.17	9.30
20	30	31.8	−2.0	1.36	61.61	−0.26	68.50	−6.89	9.45

Note that $2S_D = 2\sqrt{\text{Var}\ (Y-Y_p)}$ is calculated by using the coding units from the initial experiment.

TABLE 5.16. Factor levels for the third two-level factorial design for the yield experiment

Coded units	−1	0	+1	Step size
Temperature (°C)	30	35	40	5
Concentration (mg/g)	25	30	35	5

$X_1 = (T-35)/5$ and $X_2 = (C-30)/5$.

the slopes. Factor levels are defined in Table 5.16. The results of this factorial experiment are given in Table 5.17.

TABLE 5.17. Results from the third two-level factorial design for the yield experiment

Run	T	C	X_1	X_2	Y
21	40	35	+1	+1	69.06
22	40	25	+1	−1	61.00
23	30	35	−1	+1	61.81
24	30	25	−1	−1	57.48

Average = 62.34

The slope estimates from the third factorial design are given in Table 5.18. The new prediction equation is

$$Y_P = 62.34 + 2.69X_1 + 3.10X_2.$$

Again the step size must be recalculated in Table 5.19.

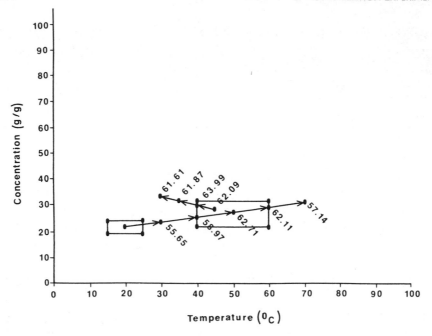

Figure 5.4.   Second steepest ascent path for yield experiment.

TABLE 5.18. Slope estimates from the third factorial design for the yield experiment

Response	T	C
High level	130.06	130.87
Low level	119.29	118.48
Difference	10.77	12.39
No. of points	4	4
Slopes	2.69	3.10

TABLE 5.19. Third step-size calculations for steepest ascent for the yield experiment

Calculations	T	C
One coded unit	5	
Desired change	5	
Change in coded units	1.0	
Slopes	2.69	3.10
Constant $R = 2.69/2(1)$	1.345	
Step size	5	5.8

The next set of steepest ascent experiments had the results given in Table 5.20.

TABLE 5.20. Third steepest ascent path for the yield experiment

Run	T	C	X_1	X_2	Y	ΔY (1.65)	Y_p	D (Y-Y_p)	$2S_D$
Base	35	30							
Step	5	5.8							
25	40	35.8	1	1.16	68.61		68.63	−0.02	13.29
26	45	41.6	2	2.32	74.55	5.94	74.91	−0.36	17.46
27	50	47.4	3	3.48	79.45	4.90	81.20	−1.75	21.70
28	55	53.2	4	4.64	83.97	4.52	87.48	−3.51	25.97
29	60	59.0	5	5.80	87.10	3.13	93.77	−6.67	30.26
30	65	64.8	6	6.96	91.00	3.90	100.06	−9.06	34.56
31	70	70.6	7	8.12	94.77	3.77	106.34	−11.57	38.86
32	75	76.4	8	9.28	95.79	1.02	112.63	−16.83	43.17
33	80	82.2	9	10.44	95.02	−0.77	118.91	−23.89	47.49

This sequence of experiments was terminated because responses in runs 32 and 33 had little change (rule 2). This result indicated that a stationary point had been reached and that either a second-order design should be run or this final position accepted as a reasonable optimum. Using steepest ascent, we were able to improve the yield to 95.8 mg/g in 33 experimental runs. Figure 5.5 shows this final steepest ascent path.

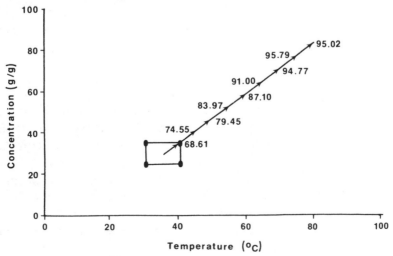

Figure 5.5.   Final steepest ascent path for yield experiment.

Now compare the steepest ascent results with the results of a one-factor-at-a-time experiment. In such an experiment, one factor is increased until the response decreases; then the other factor is increased. The optimum is the largest response. Two paths will be used: the first path varies temperature then concentration; the second path varies concentration then temperature. The initial experiment is (20°C, 20 mg/g) with a step size of 5 for both units. The results in Table 5.21 are observed.

TABLE 5.21. One-factor-at-a-time search for the yield experiment

	Path 1				Path 2		
Run	T	C	Y	Run	T	C	Y
1	20	20	49.98	1	20	20	50.08
2	25	20	50.48	2	20	25	51.48
3	30	20	54.06	3	20	30	52.81
4	35	20	54.80	4	20	35	55.22
5	40	20	55.80	5	20	40	55.40
6	45	20	55.70	6	20	45	54.89
7	40	25	61.23	7	25	40	58.68
8	40	30	64.59	8	30	40	65.39
9	40	35	66.33	9	35	40	68.18
10	40	40	71.28	10	40	40	71.73
11	40	45	71.57	11	45	40	72.78
12	40	50	74.16	12	50	40	75.16
13	40	55	74.50	13	55	40	74.14
14	40	60	77.79				
15	40	65	72.82				

The one-factor-at-a-time experiment improves the response to 72.8 mg/g for path 1 and to 74.1 mg/g for path 2. Although the improvements are about the same, they are at two different locations in the factor space. This means that for one-factor-at-a-time experiments the order in which factors are varied is important, because such experiments do not account for interactions among the factors. The two different paths for the one-factor-at-a-time experiment are illustrated in Figure 5.6.

This example was constructed by adding a random error to a known function to get a response.

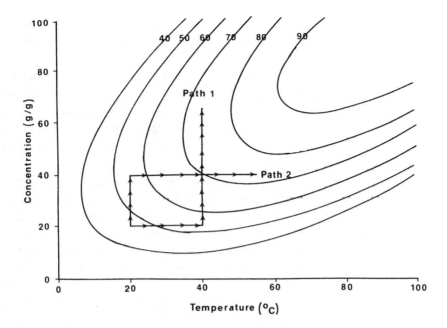

Figure 5.6.   Two paths for one-factor-at-a-time optimization.

$$Y = 20 + 0.8T + 0.8C + 0.022TC - 0.015T^2 - 0.015C^2 + \text{error}.$$

The error term added to each experimental response was taken from a table of random normal numbers (Dixon and Massey 1957) with mean zero and variance of one. The optimum value is $Y = 100$ mg/g at ($100°C$, 100 mg/g).

5.2.4   What Went Wrong

Not all results of optimization experiments meet chemists' expectations. Some possible causes of failure may be (1) near-stationary region, (2) incorrect factor selection, (3) inappropriate scaling, and (4) inappropriate model.

A near-stationary region (ie, the response region is locally nearly flat) will cause the derivatives of the first-order Taylor series to vanish or to be near zero. This does not necessarily mean that a maximum has been reached. The steepest ascent may have reached a ridge of stationary values or a minimax. A minimax, sometimes called a *saddle point*, has values that decrease in one direction and increase in another direction. The response surface may also be continuous but have step values at critical factor values. This situation may cause all first partial derivatives to be undefined at more than one point. When the nature of the response surface is the suspected cause of a problem, a second-order response surface model can be used to study the surface (Chapter 6).

The success or failure of many experiments hinges on identifying the correct factors. Sometimes seemingly unimportant factors, which are not accounted for or controlled during the experiment, can completely disrupt and negate previous findings. Careful experimental planning and the use of screening experiments to select important factors can minimize this problem.

In practice, chemists choose step sizes and scaling based on the judgment that the relative amounts will have equal affect on the response. After the initial experiment and the first steepest ascent path are carried out, it may occur that coefficients of some factors are small. This may indicate that these factors are without effect or are near their optimum value. Another possibility, however, is that the levels and scaling units adopted are disproportionately small. To remedy the situation, one should shift factor levels and increase step sizes in subsequent experiments. If the factors are without effect, the coefficients will continue to be small. If the original scales were too small, the new coefficients will be larger.

When a first-order model has a significant lack of fit, this means that a plane cannot adequately approximate the response surface in that subregion of the factor space. Interaction of the factors and curvature may be responsible for this lack of fit. One remedy is to redesign an initial experiment in a smaller region of the factor space (ie, reduce factor levels and step sizes) and proceed with steepest ascent. Another option is to augment the initial design so that a higher-order response surface can be approximated.

5.3   Simplex Optimization

Since the introduction of simplex optimization by Spendley, Hext, and Himsworth (1962), it has been applied successfully to many chemical problems (Deming and Parker, 1978). This technique is based on the geometrical figure of a simplex, which is a triangle in two dimensions, a tetrahedron in three dimensions, and a figure with $K+1$ vertices in K dimensions. A simplex optimization to maximize a response depending on K factors starts with an initial simplex design of $K+1$ experiments. To move towards the optimum measurement, a new simplex is formed by discarding the worst experimental value and replacing it with a new value in the opposite direction. Because a new simplex can be formed by adding one new experimental point, optimization can be easily adapted to the laboratory environment.

5.3.1 Fixed-Step Simplex

Assume that the proper response, factors, factor levels, and step sizes have been selected. To use simplex optimization, we define an initial simplex design by using Table 5.2. This table specifies the fraction of step sizes added to the base level and defines the vertices of the initial simplex design. After the initial simplex has been run, a new experiment is run at a vertex opposite the vertex with the worst response (ie, either lowest response for maximization or highest response for minimization). The factor levels for the new experiment and subsequent experiments are calculated from three easily applied rules.

Rule 1. To calculate new factor levels take twice the average of factor levels for retained vertices minus the factor levels for the rejected vertex.

Rule 2. Whenever the new vertex has the worst response, move out of the simplex by rejecting the vertex with the second worst response. This rule prevents a new simplex from reflecting back to the old simplex.

Rule 3. Replace old vertices. If a vertex has occurred in $K + 1$ successive simplexes and has not been eliminated, replace it with a repeated observation before proceeding. This rule prevents the system of simplexes from being anchored to some spuriously high (low) result.

To illustrate these three rules, assume a response is to be maximized that depends on temperature and the amount of acid added to a mixture. Begin by selecting an initial simplex from Table 5.2, and apply Rule 1.

TABLE 5.22. Calculation of a new simplex vertex

Vertex	Temperature (10°C)	Acid (5 ml)	Yield (%)
1	100	25.0	74
2	110	25.0	78
3	105	29.4	80
Sum of best (2 & 3)	215	54.4	
Average of best	107.5	27.2	
2 × Average of best	215	54.4	
Minus worst point (1)	−100	−25.0	
New vertex 4	115	29.4	

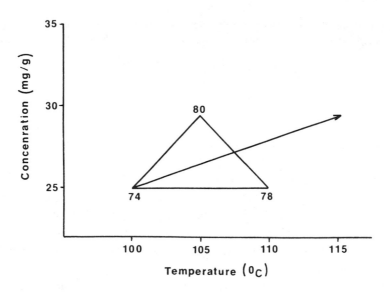

Figure 5.7.   Simplex optimization moves in a direction opposite the worst point.

Rule 1.   2 × (average of best) − worst. Table 5.22 and Figure 5.7 illustrate the application of Rule 1.

Rule 2.   Reject second worst vertex to prevent reflection to an old vertex. Table 5.23 and Figure 5.8 illustrate the application of Rule 2.

TABLE 5.23. Calculation to prevent reflection to old vertex

Vertex	Temperature (°C)	Acid (ml)	Yield (%)
2	110	25.0	78
3	105	29.4	80
4	115	29.4	76
2 × Average of 3 & 4	220	58.8	
Minus vertex 2	−110	−25.0	
New vertex 5	110	33.8	

Rule 3.   Replace or confirm an old vertex. Table 5.24 shows the need to rerun vertex 3 in this experiment.

Vertex 3 has occurred in three $K+1$ successive simplexes and still has the largest yield. This yield should be confirmed with a new observation by rerunning vertex 3 before continuing simplex optimization. If the original

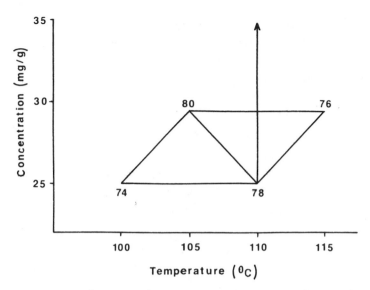

Figure 5.8.   Reject the second worst point to prevent reflection back to the previous simplex.

Table 5.24.   Replace or confirm an old vertex

Vertex	Temperature (°C)	Acid (ml)	Yield (%)
3	105	29.4	80
4	115	29.4	76
5	110	33.8	78

yield was spuriously high (due to random error), the new yield will be lower and will allow the simplex to walk in a different direction. If the original yield is a maximum value in this region, the simplex will continue to rotate around vertex 3 until it reaches the initial experiment. At this point, vertex 3 may be considered optimal, and some confirmatory runs may be made. Otherwise, the experiment should be reevaluated, and a new base level with a change of step size should be considered.

To apply simplex optimization, use only the relative ranking, not magnitude, of responses. Therefore, simplex optimization can be applied to any response that can be ranked but not necessarily measured, such as color, appearance, taste, and so on. By using the desirability coefficient (see ''Multiple Responses'' in Chapter 2), simplex optimization can also be used for multiple responses.

Simplex optimization is computationally simple; however, hand calculations frequently lead to mistakes. Using worksheets helps to avoid this
problem. For example, the worksheet in Table 5.25 can be used for the
illustrated example.

TABLE 5.25. Simplex optimization worksheet

Vertex	Temperature (10°C)	Acid (5 ml)	Yield (%)	Rank	Age
1	100	25.0	74	3	1
2	110	25.0	78	2	1
3	105	29.4	80	1	1
Sum of best	215	54.4			
Ave of best	107.5	27.2			
2 × Average	215	54.4			
Minus Worst	−100	−25.0			
4	115	29.4	76	3	1
2	110	25.0	78	2	2
3	105	29.4	80	1	2
Sum of best	220	58.8			
Ave of best	110	29.4			
2 × Average	220	58.8			
Minus Worst	−110	−25.0			
5	110	33.8	77	2	1
4	115	29.4	76	3	2
3	105	29.4	80	1	3

After the initial simplex, rule 2 is applied to subsequent simplexes if age =
1 and response has rank = $K+1$. Rule 3 applies if age = $K+1$ and rank
$\neq K+1$.

The performance of simplex optimization may be illustrated by applying
the technique to the yield experiment used to illustrate steepest ascent (see
"Steepest Ascent Example" in Chapter 5). Base level values are selected
for temperature and concentration to be (20°C, 20 mg/g) with step sizes of
5°C and 5 mg/g. Yield values are calculated by adding a standard normal
random error to the known function for each experiment. The progress of
simplexes is shown in Figure 5.9, and information on yield values is
summarized in Table 5.26. Rule 3 is applied at those vertices in Table 5.26

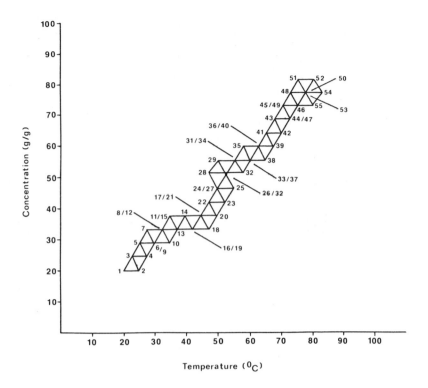

Figure 5.9.   Fixed-step-size simplex optimization of the yield experiment.

having two yield values. Rule 2 is applied in simplexes 13, 27, 34, 39, and 40 at corresponding vertices 18, 38, 46, 54, and 55. In two simplexes (ie, 13 and 34), both rules are applied. Simplex optimization stopped after 56 experiments at factor levels of (77.5°C, 77.2 mg/g) with a yield of 97.7 mg/g. This result is similar to that of steepest ascent [ie, (75°C, 76.4 mg/g) with a yield of 95.8 mg/g] but required a greater number of experimental runs.

Changing only one vertex at a time may become tedious for experiments with many factors. For these cases, more than one vertex can be discarded. Rule 1 is modified as

$$2 \times \text{(average of best points)} - \text{worst } points.$$

This modification means that several new vertices are reflections of several worst vertices. Suppose a normalized response yield depends on the factors temperature and concentration. Then Table 5.27 shows how a simplex with two new vertices is defined.

For two factors this method would not be needed, but for experiments

TABLE 5.26. Simplex optimization of the yield experiment in "Steepest Ascent Example" in Chapter 5

Vertex	Simplex	Temp	Conc	Yield
1	1	20.0	20.0	47.84
2	1,2	25.0	20.0	52.36
3	1,2,3	22.5	24.4	53.79
4	2,3,4	27.5	24.4	54.37
5	3,4,5	25.0	28.8	55.12
6/9	4,5,6,7	30.0	28.8	60.49/60.41
7	5,6	27.5	33.2	59.69
8/12	6,7,8,9	32.5	33.2	63.61/64.84
10	7,8	35.0	28.8	63.01
11/15	8,9,10,11	37.5	33.2	66.44/66.53
13	9,10	35.0	37.6	65.98
14	10,11,12	40.0	37.6	69.11
16/19	11,12,13	42.5	33.2	69.24/67.72
17/21	12,13,14,15	45.0	37.6	71.15/70.52
18	13,14	47.5	33.2	68.31
20	14,15,16	50.0	37.6	71.81
22	15,16,17	47.5	42.0	74.90
23	16,17,18	52.5	42.0	77.09
24/27	17,18,19,20	50.0	46.4	78.61/78.81
25	18,19	55.0	46.4	77.88
26/30	19,20,21,22	52.5	50.8	80.83/81.12
28	20,21	47.5	50.8	79.77
29	21,22	50.0	55.2	80.55
31/34	22,23,24,25	55.0	55.2	83.85/85.25
32	23,24	57.5	50.8	83.38
33/37	24,25,26,27	60.0	55.2	85.74/86.20
35	25,26	57.5	59.6	85.32
36/40	26,27,28,29	62.5	59.6	88.30/87.10
38	27,28	65.0	55.2	84.81
39	28,29,30	67.5	59.6	89.45
41	29,30,31	65.0	64.0	89.69
42	30,31,32	70.0	64.0	90.90
43	31,32,33	67.5	68.4	91.02
44/47	32,33,34	72.5	68.4	93.59/92.31
45/49	33,34,35	70.0	72.8	93.50/93.50
46	34,35,36	75.0	72.8	93.28
48	35,36,37	72.5	77.2	94.16
50/53	36,37,38,39,40,41	77.5	77.2	96.47/97.70
51	37,38	75.0	81.6	95.16
52	38,39	80.0	81.6	97.01
54	39,40	82.5	77.2	96.45
55	40,41	80.0	72.8	94.88
56	41	75.0	72.8	93.30

TABLE 5.27. Calculation to discard more than one vertex

Vertex	Temp	Conc	Yield
1	20.0	20.0	47.8
2	25.0	20.0	52.4
3	22.5	24.4	53.8
2 × Ave of best (3)	45.0	48.8	
Minus worst point (1)	− 20.0	− 20.0	
New vertex 4	25.0	28.8	
2 × Ave of best (3)	45.0	48.8	
Minus worst point (2)	− 25.0	− 20.0	
New vertex 5	20.0	28.8	

with three or more factors moving more than one point can be an advantage. A problem arises as to how to decide which are the best points and which are the worst. One criterion is to make two groups of responses, so that the

TABLE 5.28. Example of maximizing the response difference between the best and worst groups of experimental responses. Responses above each line are in the worst group, and those below are in the best group.

Rank	Response	Average of best	Average of worst	Difference
1	75.4			
		79.4	75.4	4.0
2	76.1			
		79.9	75.8	4.1
3	79.2			
		80.1	76.9	3.2
4	79.5			
		80.2	77.6	2.6
5	79.6			
		80.4	78.0	2.4
6	80.1			
		80.6	78.3	2.2
7	80.5			
		80.6	78.6	2.0
8	80.6			

difference between the average response of each group is a maximum. To find these groups, rank responses from lowest to highest. Next divide the responses into two groups, with the lowest values in one group and the highest values in the second group. This division can be done K ways for K factors. For each division, calculate the average response for each group and take the difference of the responses. The division that has the largest difference will indicate the best and worst vertices. For example, suppose an initial seven-factor simplex design in eight runs is used to optimize a response. Table 5.28 represents the method for determining the best and worst points. Table 5.28 shows that the best partitioning is to group those vertices associated with the two lowest responses as the worst vertices.

5.3.2   Variable Step-Size Simplex

The basic simplex method (BSM) described in the previous section does not provide for increasing or decreasing the step size. This limitation may be overcome by restarting with a smaller or larger simplex. However, Nelder and Mead (1965) introduced a modified simplex method (MSM) that allows expansion and contraction of simplexes as they progress towards the optimum region. The movement of the simplex is governed by the same rules as BSM but has additional rules for deciding what operation to use.

Figure 5.10 shows a two-dimensional simplex with vertices ranked as **B** for best, **W** for worst, and **N** for next to the worst. The vertex **R** represents the reflection of **W** through the centroid, **C**. Vector notation is used to

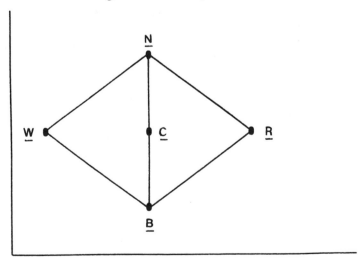

Figure 5.10.   Reflection of the worst (**W**) vertex through the centroid (**C**) to the next vertex (**R**). Vertex **B** is the best vertex, and vertex **N** is the next-to-worst vertex.

generalize the MSM to higher dimensions. Vertex **R** is found by adding the distance from **W** to **C** (ie, **C**−**W**) to the centroid:

$$\mathbf{R} = \mathbf{C} + (\mathbf{C}-\mathbf{W}).$$

This method of finding the reflection point is equivalent to rule 1 for the BSM. Three conditions can occur, depending on the rank of **R** in relation to **B** and **N**. If **R** > **B**, the new simplex should be expanded. If **N** < **R** < **B**, the new simplex should be **BNR**. If **R** < **N**, the new simplex should be contracted.

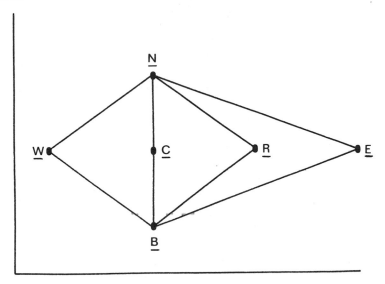

Figure 5.11.   Expansion to vertex **E** for the modified simplex method.

Expansion is shown in Figure 5.11 with **E** representing the expansion vertex. Vertex **E** is found by adding a multiple of the distance **C**− **W** to the centroid:

$$\mathbf{E} = \mathbf{C} + a(\mathbf{C}-\mathbf{W}) \quad \text{with } a > 1.$$

The value of $a = 2$ has been used in most literature examples, but this value may be modified, depending on the situation. For responses **E** > **B**, the new simplex is **BNE**; otherwise it is **BNR**, and the simplex algorithm is restarted.

Contraction (ie, **R** < **N**) is shown in Figure 5.12 for two possible situations. For both cases, only a fraction of **C**−**W** is added or subtracted to the centroid:

1. If **R** > **W**, the new simplex is **BND**1.

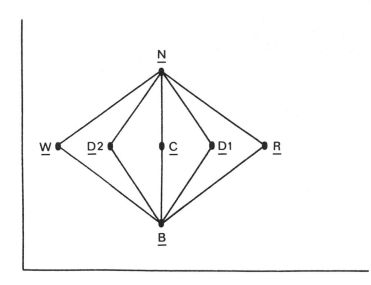

Figure 5.12.   Contraction to vertex **D**1 or vertex **D**2 for the modified simplex method.

$$\mathbf{D}1 = \mathbf{C} + b(\mathbf{C}-\mathbf{W}) \quad \text{with } 0 < b < 1.$$

2. If $\mathbf{R} < \mathbf{W}$, the new simplex is **BND2**.

$$\mathbf{D}2 = \mathbf{C} - b(\mathbf{C}-\mathbf{W}) \quad \text{with } 0 < b\, , < 1.$$

The usual value for the contraction constant is $b = 0.5$, but this value can be chosen to fit the situation. Contraction will fail for responses $\mathbf{D}1 < \mathbf{R}$ or for responses $\mathbf{D}2 < \mathbf{W}$. If the contraction vertex is the worst vertex in the new simplex, do not reject that vertex but rather reject the next to the worst vertex $\mathbf{N}$. A flowchart of MSM is given in Figure 5.13, showing the various decision rules.

Suppose yield is to be optimized for the factors temperature, concentration, and pH. Table 5.29 illustrates the calculations required for the initial move by using the MSM.

The centroid is opposite the worst vertex (ie, vertex 1) and is calculated by averaging the factor levels of the remaining vertices (ie, 2, 3, and 4). The distance from the centroid to the worst vertex is found by subtracting the factor levels of the worst vertex from the centroid factor levels. The reflection vertex is found by adding $\mathbf{C}-\mathbf{W}$ to the centroid. Notice that $\mathbf{R}$ is equivalent to "twice the average of the best (ie, centroid) minus the worst." For a fixed-size simplex, the calculations finish at this point. For MSM the expansion or contraction points are calculated, depending on the ranking of

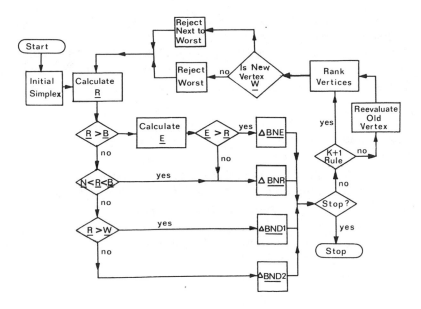

Figure 5.13.   Modified simplex method flowchart.

the response at **R**. These points are calculated by adding or subtracting a multiple of **C−W** to the centroid.

Stopping MSM depends on three conditions:

1. Preset number of vertices is exceeded.

TABLE 5.29. Calculation for modified simplex method

Vertex	Temperature (10°C)	Concentration (20 mg/g)	pH (0.5)	Yield (%)	Rank
1	80	100	5.5	50	4
2	90	100	5.5	55	3
3	85	117	5.5	60	2
4	85	106	5.9	65	1
C: centroid	86.7	107.7	5.6		
(**C−W**)	6.7	7.7	0.1		
R	93.4	115.4	5.7		
C + 2.0 (**C−W**)	100.1	123.1	5.8		
C + 0.5 (**C−W**)	90.1	111.6	5.65		
C − 0.5 (**C−W**)	83.4	103.9	5.55		

2. Preset distance between vertices is too small.

3. Response variance is too small.

Before optimization is started, the chemist must determine the number of experiments that can be accomplished by considering cost, time, and other constraints. These considerations determine an upper limit on the number of simplex vertices.

As variable-size simplexes approach the optimum, the size will contract to a smaller and smaller simplex. A criterion for stopping the search is to place a lower limit on the change of simplex coordinates to a preset value. This preset value can be formulated as a fraction of the original step size [eg, lower limit = 0.1(initial step size)]. For example, let a new vertex, $\mathbf{R}$, with coordinates $(r_1, \ldots, r_K)$ be a reflection of a worst vertex, $\mathbf{W}$, with coordinates $(w_1, \ldots, w_K)$. Step sizes for each factor are denoted by S_1, $\ldots, S_K$. The search is terminated if coordinate changes are less than a fraction, f, of the step sizes:

$$|r_1 - w_1| < fS_1,$$

$$\cdot$$
$$\cdot$$
$$\cdot$$

$$|r_K - w_K| < fS_K.$$

Responses will not usually increase substantially as the search nears the optimum. Small response variations may also occur if the simplex is on a response surface plateau. For either event, the search does not provide sufficient gain and should be terminated for evaluation. A small response variation means different response values at the vertices are due only to random error. The sample variance of the $K+1$ responses in the simplex, S_S^2, with $V1 = K$ degrees of freedom is used as a measure of response variation due to the simplex. Let S^2 with $V2$ degrees of freedom be a previous estimate of response variance from replicate experimental units. When response variation is due just to random error, both variance estimates have the same expected value, and their ratio value should be near one. This ratio of variance estimates is really a random variable. If we assume that both estimates are estimating the random error variance, this random variable has an F-distribution, $\Pr(S_S^2/S^2 < F) = p$. Therefore, the research is terminated for a probability value p if

$$S_S^2 < FS^2.$$

TABLE 5.30. F-values for the ratio of simplex variance estimate (df = $V1$) to an independent variance estimate (df = $V2$)

$$Pr(S_s^2/S^2 < F) = p$$

Denominator df ($V2$)	p	Numerator df ($V1$)				
		2	3	4	5	6
2	0.50	1.00	1.14	1.21	1.25	1.28
	0.67	2.00	2.15	2.22	2.27	2.30
	0.75	3.00	3.15	3.23	3.28	3.31
3	0.50	0.88	1.00	1.06	1.10	1.13
	0.67	1.62	1.72	1.77	1.79	1.82
	0.75	2.28	2.36	2.39	2.41	2.42
4	0.50	0.83	0.94	1.00	1.04	1.06
	0.67	1.46	1.55	1.59	1.61	1.62
	0.75	2.00	2.05	2.06	2.07	2.08
5	0.50	0.80	0.91	0.96	1.00	1.02
	0.67	1.38	1.45	1.48	1.50	1.51
	0.75	1.85	1.88	1.89	1.89	1.89
6	0.50	0.78	0.89	0.94	0.98	1.00
	0.67	1.33	1.39	1.42	1.43	1.44
	0.75	1.76	1.78	1.79	1.79	1.78

Table 5.30 gives F-values for probability values of $p = 0.50, 0.67$, and 0.75. If $p = 0.75$ is chosen, there is a probability of 0.25 that the search will continue when it should have been stopped. Note that significant levels are larger than the usual 0.05 significant level used for hypothesis testing. The reason for this modification is that it is better to continue doing some redundant experiments than to stop prematurely.

The MSM will now be illustrated for the yield experiment in Chapter 5 in the "Steepest Ascent Example" subsection. Simplex progress is shown in Figure 5.14, and numerical values are given in Table 5.31. The search was stopped after 20 experimental runs at temperature and concentration of (93.37°C, 101.30 mg/g). The yield value at this location is 98.97 mg/g, which is close to the true maximum of 100 mg/g. Preset criteria to stop the search were (1) number of vertices < 56 (ie, the number used for BSM), (2) differences between factor values for reflection vertex and previous worst

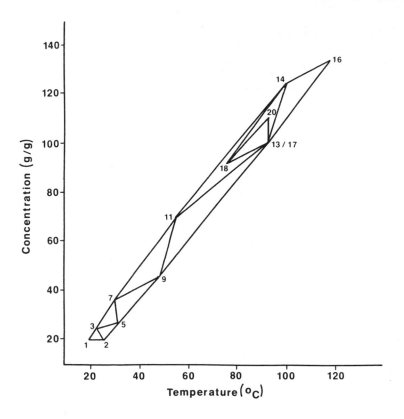

Figure 5.14.   Modified simplex method optimization of the yield experiment.

vertex must be greater than 0.1 of their corresponding factor step size, and (3) variance of responses for a simplex must be greater than $(2.00)(2.94) = 5.88$. For this last condition, $F = 2.00$ is the F-value in Table 5.30 for $p = 0.67$ with $V1 = 2$ and $V2 = 2$, and $S^2 = 2.94$ is the variance estimate made from three center points in the initial experiment of the steepest ascent example in Chapter 5. Criteria 1 and 2 did not affect the stopping of the search. The search was stopped because in simplex 10 the variance estimate $S_S^2 = 5.41$ is less than the preset limit.

The new vertices in simplexes 1 to 6 are expansion vertices. Reflection vertices are the new vertices in simplexes 7 and 9. The vertices were contracted (using $b = -0.5$) in simplexes 8 and 10. At vertex 19 the distance between best centroid and worst point is negative (ie, temp $= -16.00$, and conc $= -29.17$). The MSM rules indicate the new vertex should be contracted with $b = -0.5$, so that the new vertex is

TABLE 5.31. Modified simplex method for the yield experiment in Chapter 5. Symbols E (expansion), R (reflection), and D2 (contraction, $b = -0.5$) identify vertex types

Vertex	Simplex	Temp (°C)	Conc (mg/g)	Yield (%)
1	1	20.00	20.00	47.81
2	1,2	25.00	20.00	52.59
3	1,2,3	22.50	24.40	54.06
4	R	27.50	24.40	55.52
5E	2,3,4	31.25	26.60	58.12
6	R	28.76	31.00	59.84
7E	3,4,5	30.64	36.50	62.71
8	R	39.40	38.70	71.27
9E	4,5,6	47.85	45.85	75.94
10	R	47.25	55.76	79.74
11E	5,6,7	55.25	70.34	85.50
12	R	72.46	79.70	95.57
13	6,7,8,9,10	93.37	101.30	99.38
14R	7,8,9	100.77	125.79	91.72
15	R	138.89	156.76	76.62
16D2	8	117.98	135.16	90.29
17E	6,7,8,9,10	93.37	101.30	98.55
18R	9,10	76.16	91.94	94.44
19	R	68.77	67.45	89.01
20D2	10	92.77	111.21	95.79

$$T = 84.77 - 0.5(-16.00) = 92.77,$$
$$C = 96.62 - 0.5(-29.17) = 111.21.$$

Note that vertex 17 was run to confirm the response at vertex 13. This confirmatory experiment is an application of the $K+1$ rule used for BSM.

The *supermodified simplex method* (Routh et al, 1977) and *weighted centroid method* (Ryan et al, 1980) are additional modifications of MSM. These modifications are gradient methods that use the response magnitudes to find a new vertex. A study (Ryan et al, 1980) comparing these modifications with MSM showed that MSM efficiently minimized (optimized) a variety of functions, from any starting point, with any initial simplex size. The number of runs compared favorably with the modified gradient methods. The authors postulated that the modified methods do indeed find the neighborhood of the minimum more rapidly than MSM but are less efficient when forced to determine the minimum point itself. Near the minimum of a function, the gradient becomes poorly defined, and at the minimum there is no gradient.

6

Response Surfaces

Fitting a response surface may be considered a second stage of optimization experiments, to follow steepest ascent or simplex optimization. A response surface, used for approximating the surface in the optimum region, characterizes a response that depends on K factors by a surface in $K + 1$ dimensions. If the domain of interest is small, a surface can usually be approximated by a plane or first-order polynomial. Larger regions of interest may have to account for curvature, and a second-order polynomial is frequently a good approximation. Curvature is often detected during optimization experiments.

Chapter 6 shows how to construct central composite designs to estimate coefficients for a second-order model. These designs have several desirable properties: (1) all coefficients of a second-order model can be estimated; (2) lack of fit can be detected; (3) blocking can be used; (4) designs can be constructed to satisfy variance criteria such as rotatability; and (5) central composite designs can be constructed by augmenting first-order designs.

6.1 Second-Order Model

Two goals of response surface methodology are to find an approximating function for predicting future responses, and to determine factor values that optimize the response function. To approximate a response surface, we assume a function exists that relates the response to K factors. This true function is usually unknown and may be complicated, involving both quantitative and qualitative factors. However, a low-order polynomial can frequently be used to approximate the true function in the region of interest.

A second-order approximating model for an expected response, $E(Y)$, has the form

$$E(Y) = B_0 + \sum_{h=1}^{K} B_h X_h + \sum_{h=1}^{K} B_{hh} X_h^2 + \sum_{h=1}^{K-1} \sum_{j=h+1}^{K} B_{hj} X_h X_j.$$

There are three types of coefficients for K factors, $X_1, \ldots, X_K$: first-order coefficients, B_h's; pure quadratic coefficients, B_{hh}'s; and mixed second-order coefficients, B_{hj}'s. Therefore, including the intercept, there are $1 + 2K + K(K-1)/2$ parameters to be estimated in a second-order model.

Second-order equations can closely approximate a variety of surfaces. This flexibility is demonstrated by first reducing the second-order equation to its *canonical form*. The canonical form is found by the analytical geometry techniques of translation (to remove first-order terms) and rotation (to remove mixed second-order terms). The canonical form consists of a constant and pure quadratic terms:

$$E(Y) = C_0 + C_1 W_1^2 + C_2 W_2^2 + \cdots + C_K W_K^2.$$

The origin has been translated to the center of the response system, and the new coordinate axes $W_1, \ldots, W_K$ are the orthogonal principal axes of the contour system. For example, suppose a second-order model in two factors is used to represent the expected response

$$E(Y) = B_o + B_1 X_1 + B_2 X_2 + B_{11} X_1^2 + B_{22} X_2^2 + B_{12} X_1 X_2.$$

The type of conic or general shape of the response surface can be determined by its discriminant, and these types are summarized in Table 6.1.

TABLE 6.1. Types of conics for a two-factor second-order model

Discriminant $B_{12}^2 - 4B_{11}B_{22}$	Conic	Degenerate cases
Zero	Parabola	Parallel lines
Negative	Ellipse	Point-ellipse
Positive	Hyperbola	Intersecting lines

Rotation of the axes through an angle θ ($0 \leq \theta \leq \pi/2$ radians) is done by making the following transformation:
1. $\tan(2\theta) = B_{12}/(B_{11} - B_{22})$.
2. $\cos(2\theta) = \pm 1/\sqrt{1 + \tan^2(2\theta)}$, where $\cos(2\theta)$ has the same sign as $\tan(2\theta)$.

3. $\sin(\theta) = +\sqrt{[1-\cos(2\theta)]/2}$, and $\cos(\theta) = +\sqrt{[1+\cos(2\theta)]/2}$.
4. $X_1 = W_1\cos(\theta) - W_2\sin(\theta)$, and $X_2 = W_1\sin(\theta) + W_2\cos(\theta)$.

These transformations for X_1 and X_2 will remove cross-product terms. Linear terms are removed by translating the new factors W_1 and W_2. For example, the yield experiment example in Chapter 5 was generated by using the equation

$$E(Y) = 20 + 0.8\ X_1 + 0.8\ X_2 - 0.015\ X_1{}^2 - 0.015\ X_2{}^2 + 0.022\ X_1X_2.$$

Discriminant $= (0.022)^2 - 4(-0.015)(-0.015) = -0.000416$. Therefore, the equation is an ellipse. The axes are rotated through an angle θ such that

$$\tan(2\theta) = \frac{0.022}{-0.015 - (-0.015)} = \infty;$$

therefore, $\theta = \pi/4$. Hence, $\cos(\theta)$ and $\sin(\theta)$ both have values of $\sqrt{2}/2$, and by substituting

$$X_1 = \frac{\sqrt{2}}{2}\ W_1 - \frac{\sqrt{2}}{2}\ W_2 \text{ and } X_2 = \frac{\sqrt{2}}{2}\ W_1 + \frac{\sqrt{2}}{2}\ W_2$$

into the ellipse, we get the equation

$$E(Y) = 20 - 0.8\sqrt{2}\ W_1 - 0.004\ W_1{}^2 - 0.026\ W_2{}^2.$$

To remove first-order terms, complete the square for W_1 terms:

$$E(Y) = 100 - 0.004(W_1 - \sqrt{2} \times 100)^2 - 0.026\ W_2{}^2;$$

then substitute $U_1 = W_1 - \sqrt{2} \times 100$ and $U_2 = W_2$ for the final form

$$E(Y) = 100 - 0.004\ U_1{}^2 - 0.026\ U_2{}^2.$$

A general method for transforming second-order equations in K factors to their canonical form is given in Myers (1971).

In Figure 6.1, contours for the canonical form

$$E(Y) = 10 + B_{11}\ X_1{}^2 + B_{22}\ X_2{}^2,$$

are plotted for coefficient values of -1, 0, and $+1$. Three types of systems can be distinguished, which represent a maximum, a stationary ridge, and a minimax. Type 6.1(a) illustrates a system centered on the maximum with the canonical equation having both coefficients negative. If both coefficients were positive, the system would be centered on the minimum with elliptical contours of increasing values. Type 6.1(b) illustrates a stationary ridge

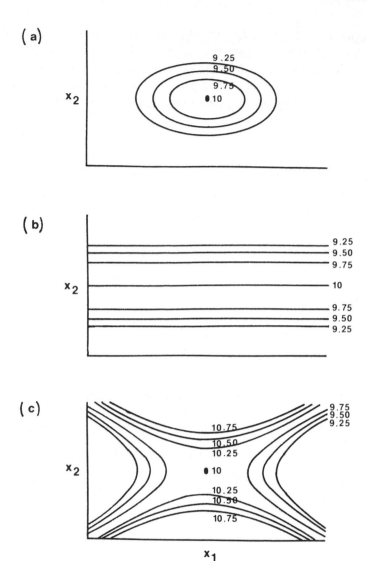

Figure 6.1. Contours for the two-factor canonical form: $E(Y) = 10 + B_{11}X_1^2 + B_{22}X_2^2$ with coefficient values of (a) $(-1, -1)$, (b) $(0, -1)$, and (c) $(-1, +1)$.

system with $B_{11} = 0$ and $B_{22} = -1$, where the contours are lines of decreasing values. There is no unique maximum because the highest response of $E(Y) = 10$ can be obtained at any point on the X_1 axis (ie, $X_2 = 0$). A stationary ridge system would also occur if $B_{11} = 0$ and $B_{22} = +1$ except that values of the contour lines would be increasing as the magnitude of $|X_2|$ increased, and the minimum value along the X_1 axis would be $E(Y)$

$= 10$. Type 6.1(c) illustrates a minimax system obtained with $B_{11} = -1$ and $B_{22} = +1$, with contours being hyperbolas. For this system, contour values decrease along the X_1 axis and increase along the X_2 axis. The system is centered on the minimax value, which is also called a *saddle point*.

Figures 6.1(a) and (c) are centered on the *stationary point*, and Figure 6.1(b) is symmetric about a *stationary ridge*. The stationary point for a response surface is the factor coordinates (or a region for a ridge system) at which derivatives are zero for the approximating function with respect to each factor. Figure 6.1(a) shows that the stationary point is not necessarily a minimum or a maximum. To derive an expression for the stationary point for a second-order response function, we use the following convenient matrix notation for the expected response:

$$E(Y) = B_0 + x'b + x'B\,x,$$

where

$$
x = \begin{bmatrix} x_1 \\ x_2 \\ \vdots \\ x_K \end{bmatrix}, \quad
b = \begin{bmatrix} B_1 \\ B_2 \\ \vdots \\ B_K \end{bmatrix}, \quad
B = \begin{bmatrix} B_{11} & B_{12}/2 & \dots & B_{1K}/2 \\ & B_{22} & \dots & B_{2K}/2 \\ & & \dots & \\ \text{sym} & & & B_{KK} \end{bmatrix}
$$

The derivatives of $E(Y)$ with respect to vector x are equated to 0 and solved for the stationary point, x, to give

$$x_S = -\frac{1}{2} B^{-1} b$$

The stationary point is estimated by substituting the estimated coefficients into b and B. For example, suppose the estimated prediction equation is

$$Y_p = 20 + 3x_1 - 2x_2 - x_1^2 + 4x_2^2 - 2x_1 x_2.$$

Then

$$
\hat{b} = \begin{bmatrix} 3 \\ -2 \end{bmatrix}, \quad
\hat{B} = \begin{bmatrix} -1 & -1 \\ -1 & 4 \end{bmatrix}.
$$

The estimated stationary point is

$$
\hat{x}_S = -\frac{1}{2} \begin{bmatrix} -0.8 & -0.2 \\ -0.2 & 0.2 \end{bmatrix} \begin{bmatrix} 3 \\ -2 \end{bmatrix} = \begin{bmatrix} 1.0 \\ 0.5 \end{bmatrix}
$$

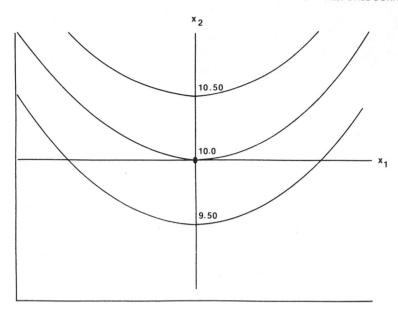

Figure 6.2. Stationary point outside the experimental region.

The coordinates of the stationary point may lie outside the experimental region defined by the factor levels, as indicated in Figure 6.2. In this figure the maximum lies in the direction of the positive X_2 axis. It is important to remember that the approximating response function is expected to be adequate only in the immediate neighborhood of the experimental design. No conclusion about the position or existence of the stationary region should be made outside the experimental region because the fitted surface is dependable only in the region of available data. However, results will indicate the direction for future experiments.

6.2   Central Composite Designs

An experimental design constructed to estimate coefficients for an approximating model should meet some or all of the following design criteria:
  1. Provide estimates for all coefficients in the approximating model.
  2. Require a small number of experimental units.
  3. Provide a test for lack of fit.
  4. Allow the experiment to be performed in blocks.

5. Allow specified variance criteria to be met for estimated coefficients and estimated responses.

Estimating coefficients for second-order terms in an approximating model requires that each factor appear at more than two levels. The most obvious design is a 3^K factorial design. This design satisfies several design criteria. However, the number of experimental units required can grow rapidly as the number of factors increases. To overcome this problem, Box and Wilson (1951) introduced *central composite designs* that do not require an excessive number of experimental units. These designs are composed of 2^K factorial designs augmented by $2K$ experimental units on the K axes equally spaced at a units and having one or more experimental units at the origin. Figure 6.3 illustrates a central composite design for two factors, and Table 6.2 gives the number of design points required for a K-factor second-order model.

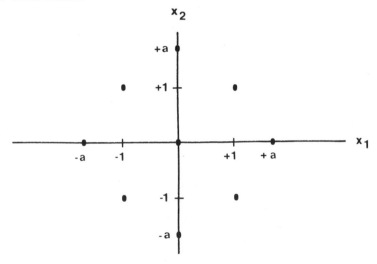

Figure 6.3. Two-factor central composite design composed of four cube points, four axial points, and one center point.

The number of experimental units for a 3^K factorial design is usually impractical for $K > 3$, although fractions of full factorials could be used. Central composite designs require fewer experimental units. In addition, for $K > 4$, half-fractions of full factorials may be used and all second-order coefficients can still be estimated. Center points are frequently replicated to allow a test for lack of fit and to permit running the experiment in blocks.

The general form of a central composite design is composed of $N_c = 2^{K-P}$ cube points (ie, factorial or fractional factorial points), $N_a = 2K$ axial

TABLE 6.2. Number of design points needed to estimate coefficients for a K-factor second-order model

K factors	Number of experimental units		Number of coefficients
	3^K factorial	Central composite	
2	9	9	6
3	27	15	10
4	81	25	15
5	243	43	21
6	729	77	28

points, and N_0 center points for a total of $N = N_c + N_a + N_0$ experimental units. Table 6.3 gives coordinates in the coded factor space for the general central composite design.

TABLE 6.3. Coordinates for the general composite design for K factors

Type	Coordinates	No. of points
Cube points:	$(\pm 1, \pm 1, ..., \pm 1)$	$N_c = 2^{K-P}$
Axial points:	$(\pm a, 0, ..., 0)$ $\quad$ ... $\quad$ $(0, 0, ..., \pm a)$	$N_a = 2K$
Center points:	$(0, 0, ..., 0)$	N_o

Frequently, central composite designs are constructed by adding points to a factorial design that has been used for steepest ascent optimization. If a test for lack of fit shows that a first-order design does not approximate the response surface, the initial factorial design can be augmented by axial and center points to fit a second-order model.

To construct a central composite design, we must determine values for N_c, N_a, N_0, and "a." Values of N_c and N_a depend on the value of K, with $N_c = 2^{K-P}$ being the number of experimental units for a factorial or a fractional factorial (ie, $P \geq 1$ if $K > 4$), and $N_a = 2K$. If a fractional factorial is used, it must be of resolution V or larger to allow estimation of all second-order coefficients. The number of center points, N_0, should be greater than one to estimate pure error for a lack of fit test. How much greater than one will be determined by requirements for blocking and satisfying variance criteria for Var($\mathbf{b}$) and Var(Y_p). The value of a for axial points will also be determined by variance criteria for estimated coefficients and estimated responses.

6.2.1   Orthogonal Blocking

Chemists do not always have enough raw material or fixed periods of time to guarantee uniform experimental units for all experimental points in a central composite design. To overcome this limitation, experimental units are divided into groups or blocks that are considered homogeneous units. Blocking is frequently used to eliminate time trend effects that often inflate experimental error estimates. The effect due to blocking is assumed to add only a constant value to the response in a particular block. Central composite designs can be run in blocks, so that additive block effects are not confounded with coefficients of the second-order model. Eliminating this confounding is called *orthogonal blocking*.

To represent a block additive effect, blocking variable Z_w will have a value of $Z_w = 1$ in the wth block, and $Z_w = 0$ outside wth block. Each blocking variable is scaled by subtracting its average over all experimental units (ie, N) from each value of Z_w. This average is $\bar{Z}_w = M_w/N$ for the wth blocking variable, with M_w representing the number of experimental units in the wth block. For two blocks and two factors, the expected response is

$$E(Y) = B_0 + B_1 X_1 + B_2 X_2 + B_{11} X_1^2 + B_{22} X_2^2$$
$$+ \; C_1 (Z_1 - \bar{Z}_1) + C_2 (Z_2 - \bar{Z}_2).$$

In the model matrix the value of $Z_w - \bar{Z}_w = 1 - M_w/N$ in the wth block, and $Z_w - \bar{Z}_w = -M_w/N$ outside the wth block. The column vectors for block variables are orthogonal to the column vector for B_0, and, therefore, blocking coefficients C_w are not confounded with B_0 .

Two conditions must be satisfied for blocking coefficients not to be confounded with first- and second-order coefficients. First, in the design matrix, column vectors for blocking variables must be orthogonal to column vectors for first-order and mixed second-order terms. This condition implies that each block must be a first-order orthogonal design. Central composite designs can be blocked into first-order orthogonal designs by allocating cube points to one block and axial points to a second block. Additional blocking can be achieved by partitioning cube points into fractional factorials of resolution III.

Secondly, in the model matrix, column vectors for blocking must also be orthogonal to column vectors for pure second-order terms. This condition implies that the fraction of the total sum of squares for the hth variable ($h = 1, 2, \ldots, K$) in each block is proportional to the number of experimental units in that block:

$$\frac{\sum_{j=1}^{Mw} X_{hj}^2}{\sum_{j=1}^{N} X_{hj}^2} = \frac{M_w}{N}.$$

For two blocks let the number of center points be divided between the cube block, N_{oc}, and axial block, N_{oa} (ie, $N_o = N_{oc} + N_{oa}$). Then for each block:
Cube block:

$$\sum_{j=1}^{Mc} X_{hj}^2 = N_c, \quad \text{and} \quad M_c = N_c + N_{oc}.$$

Axial block:

$$\sum_{j=1}^{Ma} X_{hj}^2 = 2a^2, \quad \text{and} \quad M_a = 2K + N_{oa}.$$

The ratio of axial block points to cube block points gives

$$\frac{2a^2}{N_c} = \frac{2K + N_{oa}}{N_c + N_{oc}},$$

or

$$a^2 = \frac{N_c (2K + N_{oa})}{2(N_c + N_{oc})}.$$

For example, a central composite design is constructed for two factors in two blocks with four center points divided equally between the two blocks.

$$N_c = 2^2 = 4, \quad N_a = 2(2) = 4, \quad N_{oc} = 2, \quad \text{and} \quad N_{oa} = 2.$$

Then for orthogonal blocking, choose

$$a = \sqrt{4(4 + 2)/(2(4 + 2))} = \sqrt{2} = 1.414.$$

Therefore, the design consists of a cube block and an axial block as

$$
\text{Cube block} = \left\{ \begin{array}{cc}
+1 & +1 \\
+1 & -1 \\
-1 & +1 \\
-1 & -1 \\
0 & 0 \\
0 & 0
\end{array} \right.
$$

$$
\text{Axial block} = \left\{ \begin{array}{cc}
+1.414 & 0 \\
-1.414 & 0 \\
0 & +1.414 \\
0 & -1.414 \\
0 & 0 \\
0 & 0
\end{array} \right.
$$

For this example, center points were divided equally between two blocks. This division is usually appropriate. However, to satisfy additional design criteria, we can make different partitions of center points. Different partitions will determine different values of axial spacing (ie, $\pm a$). In Table 6.4 the partitioning of center points is given for cube blocks and axial blocks to determine axial spacing, which nearly determines the variance criterion of rotatability. Rotatability criteria are given in the subsection "Rotatability."

If the experimental situation dictates that more than two blocks are needed, cube points can be divided into fractional factorials. For example, three blocks with $K = 3$ can be formed from a central composite design by using two half-fractions of 2^3 in two blocks defined by $\mathbf{I} = \pm \mathbf{1}*\mathbf{2}*\mathbf{3}$ and axial points in the third block. Five blocks with $K = 5$ can be formed by using four 2^{5-2} fractional factorials for four blocks, defined by $\mathbf{I} = \pm \mathbf{1}*\mathbf{2}*\mathbf{3}*\mathbf{4} = \pm \mathbf{1}*\mathbf{2}*\mathbf{5} = \pm \mathbf{3}*\mathbf{4}*\mathbf{5}$, and axial points in the fifth block. The center points would be partitioned among the blocks to get a desired value for axial spacing. Axial spacing is calculated in the same manner as the two-block case with the number of cube points equal to the number of factorial or fractional factorial points (ie, $N_c = 2^{K-P}$).

TABLE 6.4. Partition of center points for a cube block and axial block that nearly determines axial spacing for rotatability

K	N	N_c	N_a	N_{oc}	N_{oa}	Axial spacing	K	N	N_c	N_a	N_{oc}	N_{oa}	Axial spacing
2	10	4	4	1	1	1.414	5a	28	16	10	2	0	2.108
	11	4	4	1	2	1.549		29	16	10	3	0	2.052
	12	4	4	2	2	1.414		30	16	10	4	0	2.000
	13	4	4	2	3	1.528		31	16	10	5	0	1.952
	14	4	4	3	3	1.414		32	16	10	5	1	2.047
	15	4	4	3	4	1.512		33	16	10	6	1	2.000
	16	4	4	4	4	1.414		34	16	10	7	1	1.956
3	16	8	6	1	1	1.764	6	78	64	12	0	2	2.646
	17	8	6	2	1	1.673		79	64	12	0	3	2.739
	18	8	6	2	2	1.789		80	64	12	0	4	2.828
	19	8	6	3	2	1.706		81	64	12	1	4	2.807
	20	8	6	4	2	1.633		82	64	12	2	4	2.785
	21	8	6	4	3	1.732		83	64	12	2	5	2.871
	22	8	6	5	3	1.664		84	64	12	3	5	2.849
4	26	16	8	1	1	2.058	6a	46	32	12	2	0	2.376
	27	16	8	2	1	2.000		47	32	12	3	0	2.342
	28	16	8	3	1	1.947		48	32	12	3	1	2.438
	29	16	8	3	2	2.052		49	32	12	4	1	2.403
	30	16	8	4	2	2.000		50	32	12	5	1	2.371
	31	16	8	5	2	1.952		51	32	12	6	1	2.340
	32	16	8	5	3	2.047		52	32	12	6	2	2.428
5	44	32	10	0	2	2.449
	45	32	10	1	2	2.412
	46	32	10	2	2	2.376
	47	32	10	3	2	2.342
	48	32	10	3	3	2.438
	49	32	10	4	3	2.403
	50	32	10	5	3	2.371

5a, 6a: Half-fractional factorial is used for the cube block.

6.2.2 Uncorrelated Coefficient Estimates

Coefficients for second-order models are estimated by using least squares. Variances and covariances for estimated coefficients are expressed using the covariance matrix (see "Estimating the Covariance Matrix" in Chapter 4)

$$\text{Var}(\mathbf{b}) = (\mathbf{X}'\mathbf{X})^{-1}\sigma^2$$

Values on the diagonal represent the variances of estimated coefficients, and off-diagonal values represent covariances between the estimated coefficients. For a central composite design, values for N_c, N_a, N_o, and axial spacing determine the covariance matrix. Desirable properties for the covariance matrix can be determined by careful selection of these values.

A covariance matrix for a two-factor second-order model is used to illustrate the dependency on parameters of a central composite design. For this design, $N_c = 4$, $N_a = 4$ with axial spacing of $\pm a$, and N_o center points. The model matrix is

$$\mathbf{X} = \begin{bmatrix}
X_0 & X_1 & X_2 & X_1^2 & X_2^2 & X_1X_2 \\
1 & +1 & +1 & +1 & +1 & +1 \\
1 & +1 & -1 & +1 & +1 & -1 \\
1 & -1 & +1 & +1 & +1 & -1 \\
1 & -1 & -1 & +1 & +1 & +1 \\
1 & +a & 0 & a^2 & 0 & 0 \\
1 & -a & 0 & a^2 & 0 & 0 \\
1 & 0 & +a & 0 & a^2 & 0 \\
1 & 0 & -a & 0 & a^2 & 0 \\
1 & 0 & 0 & 0 & 0 & 0 \\
\vdots & \vdots & \vdots & \vdots & \vdots & \vdots \\
1 & 0 & 0 & 0 & 0 & 0
\end{bmatrix}$$

The matrix $\mathbf{X}'\mathbf{X}$ is formed in terms of $N = N_c + N_a + N_0$ and the axial spacing as

$$\mathbf{X'X} = \begin{bmatrix} N & 0 & 0 & N_c+2a^2 & N_c+2a^2 & 0 \\ 0 & N_c+2a^2 & 0 & 0 & 0 & 0 \\ 0 & 0 & N_c+2a^2 & 0 & 0 & 0 \\ N_c+2a^2 & 0 & 0 & N_c+2a^4 & N_c & 0 \\ N_c+2a^2 & 0 & 0 & N_c & N_c+2a^4 & 0 \\ 0 & 0 & 0 & 0 & 0 & N_c \end{bmatrix}$$

The inverse of this matrix is found by solving equations from the relation $(\mathbf{X'X})(\mathbf{X'X})^{-1} = \mathbf{I}$:

$$(\mathbf{X'X})^{-1} = \begin{bmatrix} A & 0 & 0 & B & B & 0 \\ 0 & C & 0 & 0 & 0 & 0 \\ 0 & 0 & C & 0 & 0 & 0 \\ B & 0 & 0 & D & E & 0 \\ B & 0 & 0 & E & D & 0 \\ 0 & 0 & 0 & 0 & 0 & F \end{bmatrix} ,$$

where, for $K = 2$, the elements are

$$Q = \frac{1}{2a^4[N(KN_c + 2a^4) - K(N_c + 2a^2)^2]} ,$$

$$A = 2a^4(KN_c + 2a^4)Q,$$

$$B = -2a^4(N_c + 2a^2)Q,$$

$$C = \frac{1}{N_c + 2a^2} ,$$

$$D = [(N_c + 2a^2)^2 - NN_c]Q + \frac{1}{2a^4} ,$$

$$E = [(N_c + 2a^2)^2 - NN_c]Q,$$

$$F = \frac{1}{N_c} .$$

Inspecting the covariance matrix shows that covariances are zero between the estimated intercept and first-order coefficients and between the estimated intercept and estimated mixed second-order coefficients.

Covariances between the estimated intercept and estimated pure second-order coefficients, $\text{Cov}(b_o, b_{jj})$, are not zero because $Q > 0$, and, therefore, $B < 0$. Covariances between estimates of pure second-order coefficients, $\text{Cov}(b_{11}, b_{22})$, are zero if

$$NN_c = (N_c + 2a^2)^2.$$

This condition for uncorrelated, estimated, pure second-order coefficients is true for any number of factors. A central composite design that satisfies this condition is called *orthogonal*, but it should not be confused with orthogonal blocking, and it does not apply to covariances between the estimated intercept and estimated second-order coefficients. The axial spacing needed for orthogonality is

$$a^2 = \frac{\sqrt{(N_c + N_a + N_0)N_c} - N_c}{2}.$$

For example, an orthogonal central composite design for two factors with $N_c = 4$, $N_a = 4$, and $N_o = 1$ has an axial spacing of $a = 1$. This design is equivalent to a 3^2 factorial design. If the number of center points is increased from $N_o = 1$ to $N_o = 2$, axial spacing is increased to $a = 1.162$.

6.2.3 Rotatability

Orthogonality is a condition that eliminates covariances between estimated pure second-order coefficients. Rather than using a criterion for individual estimated coefficients, we can use criteria based on the joint effect of all coefficients. One criterion is based on variances of estimated responses for points that are an equal distance from the design center. Designs that have a constant variance of predicted responses at all points that are equidistant from the design center are called *rotatable designs*. This criterion means that $\text{Var}(Y_p)$ depends only on distance from the design center and not on direction. Variance contours of constant values are represented by circles, spheres, or hyperspheres. Although rotatability is defined for any design, the focus here will be only on central composite designs for second-order models.

An example of a design that is not rotatable is a 3^2 factorial design that is an orthogonal central composite design with axial spacing of $a = \pm 1$. As shown in the section ''3^K Factorial Designs,'' in Chapter 4, variance of a predicted response at a point x_h is given as

$$\text{Var}(Y_{ph}) = x_h'\text{Var}(b)x_h,$$

or

$$\text{Var}(Y_{ph}) = \sigma^2\left(\frac{5}{9} - \frac{x_{1h}^2 + x_{2h}^2 - x_{1h}^4 - x_{2h}^4}{2} + \frac{(x_{1h}x_{2h})^2}{4}\right).$$

The plot in Figure 6.4 shows a standardized variance function $\sigma^{-2}\text{Var}(Y_{ph})$ versus the angle θ for points on a unit circle centered at $(0,0)$. The standardized variance has a maximum value of $5/9$ at design points $(\pm 1,0)$, and $(0, \pm 1)$, but it has smaller values for positions between design points. Therefore, all central composite designs are not rotatable even if they have the added property of orthogonality.

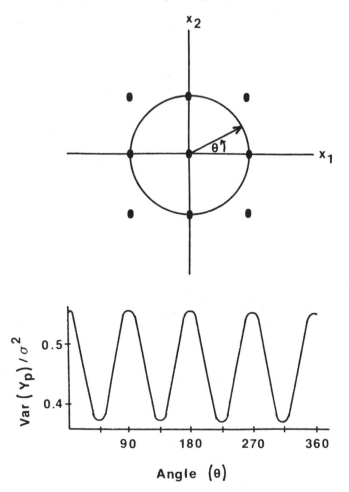

Figure 6.4. Standardized variance function for a unit circle of a 3^2 factorial design.

Box and Hunter (1957) gave a necessary and sufficient condition for a design to be rotatable. This condition implies that axial spacing for central composite designs be the fourth root of the number of cube points or

$$a^4 = N_c.$$

Rotatability does not depend on the number of center points. However, center points can be added to orthogonal designs with a corresponding increase in the axial spacing to satisfy rotatability. Table 6.5 gives the number of center points required to satisfy both rotatability and orthogonality for a central composite design.

TABLE 6.5. Number of center points required for both rotatability and orthogonality for central composite designs

K	N_c	N_a	N_o	Axial
2	4	4	8	1.414
3	8	6	9.3	1.682
4	16	8	12	2.000
5	32	10	16.6	2.378
5a	16	10	10	2.000
6	64	12	24	2.828
6a	32	12	14.6	2.378

a = one-half fractional factorial for cube points.

Table 6.5 shows two practical problems encountered for a design to satisfy both rotatability and orthogonality. First, there are designs that cannot satisfy both conditions indicated by a noninteger number of center points. Second, the number of center points required to satisfy both conditions is usually too large for many experiments. Two strategies can be adopted to compromise between rotatability and orthogonality: (1) Use axial spacing for orthogonality and add as many center points as possible to get close to rotatability; or (2) fix the axial spacing for rotatability and add as many center points as possible to get close to orthogonality.

The first strategy is demonstrated in Figure 6.5 for a two-factor central composite design. This plot is the standardized variance function on a unit circle for an increasing number of center points. Axial spacing for each additional center point is changed to maintain orthogonality. For only one

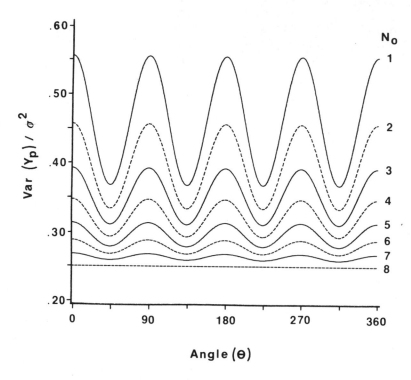

Figure 6.5. Standardized variance function for a unit circle of a two-factor central composite design for increasing number of center points (N_o). Axial spacing is changed to maintain orthogonality.

center point, the standardized variance function represents a 3^2 design and oscillates a great deal with direction. As center points are added, the oscillation is dampened until there are eight center points, and rotatability is achieved.

The second strategy is demonstrated by comparing the average value of the standardized variance function over a circle for both orthogonal designs and rotatable designs. The average standardized variance is found by integrating the standardized variance function over the angle θ and dividing by 2π. The plots in Figure 6.6 show these average values for the design center and the unit circle. For the unit circle the average standardized variance for rotatability is slightly higher than the value for orthogonality for one center point. As center points are added, average standardized variances for rotatability are less than values for orthogonality until eight center points are added, when the two values are equal. After eight center points, orthogonality values are always slightly less than those values for rotatability, but in practice more than eight center points would rarely be added.

TABLE 6.6. Differences between average standardized variances for rotatability and orthogonality at the design center

Center points	Factors						
	2	3	4	5	5a	6	6a
1	0.444	0.555	0.640	0.580	0.482	0.228	0.708
2	0.104	0.161	0.207	0.210	0.188	0.098	0.265
3	0.033	0.063	0.089	0.097	0.090	0.045	0.131
4	0.011	0.027	0.043	0.049	0.047	0.019	0.073
5	0.004	0.012	0.022	0.026	0.026	0.006	0.043

a = one-half fractional factorial used for cube points.

Average standardized variance at the design center (ie, $R = 0$) shows rotatable values larger than orthogonality values for adding up to three or four center points. With more than four center points, the average standardized variances for both rotatability and orthogonality are nearly equal. For more than two factors, Table 6.6 shows differences between average standardized variances at the design center for rotatability and orthogonality.

The second strategy for compromising between orthogonality and rotatability is better, because a smaller average standardized variance can be obtained without adding many center points. The basic recommendation would be to construct a central composite design with axial spacing satisfying rotatability and to use four center points. A summary of conditions for orthogonal blocking, orthogonality, and rotatability is given in Table 6.7.

Table 6.7 Conditions for orthogonal blocking, orthogonality, and rotatability

Design criteria	Condition
Orthogonal blocking	Form cube blocks and an axial block $$\frac{\sum_{j=1}^{M_w} X_{hj}^2}{\sum_{j=1}^{N} X_{hj}^2} = \frac{M_w}{N}$$
Orthogonality	$a^2 = (\sqrt{N N_c} - N_c)/2$
Rotatability	$a^2 = \sqrt{N_c}$

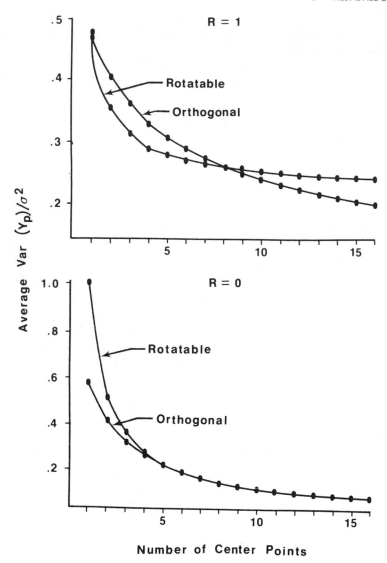

Figure 6.6. Average standardized variance function for a unit circle (R = 1) and the design center (R = 0) for a two-factor central composite design using either orthogonal or rotatable axial spacing.

6.3   Response Surface Example

Rubin et al (1971) used several response surface designs to optimize assay conditions for specific transfer ribonucleic acids (tRNA's). The reaction involved in the determination of a specific tRNA is the reversible esterification of tRNA with its corresponding L-amino acid (AA). This reaction is enzymically catalyzed by a specific aminoacyl-tRNA synthetase and is energized by adenosine triphosphate. A ^{14}C-labeled amino acid is used, and the extent of esterification (aminoacylation) is measured by the activity (counts) in the final product.

As part of this study, a central composite design was used to optimize conditions for the activity of arginine-tRNA. The three factors studied were concentration of enzyme (μg of protein), concentration of labeled amino acid (AA, pmoles), and pH of the solution. The central composite design used an axial spacing of $a = \pm1.7$, which approximates orthogonality. Six center points, $N_o = 6$, were used for a total of $N = N_c + N_a + N_o = 8 + 6 + 6 = 20$ design points. The relationship between factor levels and coded values is given in Table 6.8.

TABLE 6.8. Relationship between factor levels and coded values

	Coded values				
	−1.7	−1.0	0.0	+1.0	+1.7
Enzyme (μg protein)	3.2	6.0	10.0	14.0	16.8
Amino acid (pmoles)	860	1000	1200	1400	1540
pH	6.6	7.0	7.5	8.0	8.4

The coded values are calculated by

$$X_1 = \frac{\text{enzyme} - 10}{4},$$

$$X_2 = \frac{\text{amino acid} - 1200}{200},$$

$$X_3 = \frac{\text{pH} - 7.5}{0.5}.$$

The central composite design is given in Table 6.9.

TABLE 6.9. Central composite design for three factors

	Factors					Factors			
Trials	1	2	3	Activity	Trials	1	2	3	Activity
1	+1	+1	+1	4930	11	0	+1.7	0	4566
2	+1	+1	−1	4810	12	0	−1.7	0	4695
3	+1	−1	+1	5128	13	0	0	+1.7	4872
4	+1	−1	−1	4983	14	0	0	−1.7	4773
5	−1	+1	+1	4599	15	0	0	0	5063
6	−1	+1	−1	4599	16	0	0	0	4968
7	−1	−1	+1	4573	17	0	0	0	5035
8	−1	−1	−1	4422	18	0	0	0	5122
9	+1.7	0	0	4891	19	0	0	0	4970
10	−1.7	0	0	4704	20	0	0	0	4925

The matrix computation, described in Chapter 4, required for fitting a complete second-order model to the data can be tedious if done by hand. However, there are many computer programs that can be used for this analysis. For this example, the Statistical Analysis System (SAS, 1982) computer program was used to analyze the data. To check lack of fit for the second-order model, the residual sum of squares (SSR) is partitioned into sum of squares due to lack of fit (SSLOF) and sum of squares due to pure error (SSPE; see the subsection "Steepest Ascent Example" in Chapter 5). Pure error sum of squares is calculated from replicated center points, and SSLOF = SSR−SSPE. An F-test can be used to test the null hypothesis that the second-order model adequately fits the data. This test is done in Table 6.10 by comparing mean squares (sum of squares/degrees of freedom) of lack of fit to pure error. .

TABLE 6.10. Test for lack of fit of the second-order model

Residual	df	Sum of squares	Mean squares	F-ratio
Lack of fit	5	83262.2	16652.4	3.14
Pure error	5	26478.8	5295.8	
Residual	10	109741.0	10974.1	

The probability of getting an $F(5,5) = 3.14$ under the null hypothesis is

Pr $= 0.12$. Therefore, the second-order model is an adequate approximation to the data, and an estimation of error variance is the residual mean squares ($S^2 = 10974.1$).

Estimated variances of least squares estimates of the coefficients are calculated from $(\mathbf{X'X})^{-1} S^2$, where the inverse matrix is given Table 6.11.

TABLE 6.11. Inverse matrix, $(\mathbf{X'X})^{-1}$ for the central composite model matrix

b_0	b_1	b_2	b_3	b_{11}	b_{22}	b_{33}	b_{12}	b_{13}	b_{23}
0.167	0.000	0.000	0.000	−0.056	−0.056	−0.056	0.000	0.000	0.000
0.000	0.073	0.000	0.000	0.000	0.000	0.000	0.000	0.000	0.000
0.000	0.000	0.073	0.000	0.000	0.000	0.000	0.000	0.000	0.000
0.000	0.000	0.000	0.073	0.000	0.000	0.000	0.000	0.000	0.000
−0.056	0.000	0.000	0.000	0.067	0.007	0.007	0.000	0.000	0.000
−0.056	0.000	0.000	0.000	0.007	0.067	0.007	0.000	0.000	0.000
−0.056	0.000	0.000	0.000	0.007	0.007	0.067	0.000	0.000	0.000
0.000	0.000	0.000	0.000	0.000	0.000	0.000	0.125	0.000	0.000
0.000	0.000	0.000	0.000	0.000	0.000	0.000	0.000	0.125	0.000
0.000	0.000	0.000	0.000	0.000	0.000	0.000	0.000	0.000	0.125

Estimated variances of the coefficients derived from the inverse of the model matrix are given in Table 6.12.

TABLE 6.12. Estimated variances and covariances of least square estimated coefficients for the second-order model

$\text{Vâr}(b_0) =$	$(0.167)(10974.1) =$	1832.7
$\text{Vâr}(b_h) =$	$(0.073)(10974.1) =$	801.1
$\text{Vâr}(b_{hh}) =$	$(0.067)(10974.1) =$	735.3
$\text{Vâr}(b_{hk}) =$	$(0.125)(10974.1) =$	1371.8
$\text{Côv}(b_0,b_{hh}) =$	$(-0.056)(10974.1) =$	614.5
$\text{Côv}(b_{hh},b_{kk}) =$	$(0.007)(10974.1) =$	76.8

Although axial spacing is near the spacing needed for orthogonality, covariances between pure quadratic terms are not zero.

Least squares estimated coefficients can be tested to see if they are significantly different from zero. This test uses the t-test by comparing the t-statistic calculated as the ratio of an estimated coefficient to its standard deviation (square root of its variance). Degrees of freedom for the t-test are the degrees of freedom used to estimate S^2 (dF $= 10$, for this example). The probability of the t-statistic under the null hypothesis (see Table 3.4) is then compared to a desired significance level. Those coefficients with probabilities larger than the significance level are discarded. Remaining coefficients are recalculated for the final approximating function. Least squares estimated coefficients for a second-order model with corresponding t-statistics and probabilities are given in Table 6.13.

TABLE 6.13. Least square estimated coefficients for a second-order model with corresponding t-statistics and probabilities

Parameter	Estimate	t-Statistic	$Pr > |t|$
b_0	5013.5	117.27	0.0001
b_1	143.4	5.08	0.0005
b_2	-28.1	-1.00	0.3428
b_3	42.4	1.50	0.1639
b_{11}	-71.7	-2.64	0.0247
b_{22}	-129.5	-4.77	0.0008
b_{33}	-63.1	-2.32	0.0426
b_{12}	-71.8	-1.94	0.0815
b_{23}	14.3	0.38	0.7085
b_{23}	-22.0	-0.59	0.5657

A significance level of $\alpha = 0.10$ is used to test if estimated coefficients are significantly different from zero. This test means that if a probability in Table 6.13 is greater than 0.10, then that estimated coefficient is discarded. The significant coefficients are reestimated to give the final prediction equation

$$Y_p = 5013.5 + 143.4X_1 - 71.7X_1^2 - 129.5X_2^2 - 63.1X_3^2 - 71.8X_1X_2.$$

The inverse matrix for the reestimated prediction equation corresponds to the appropriate rows and columns in the original $(\mathbf{X'X})^{-1}$. This result occurs because the new prediction equation includes both the intercept term and all pure quadratic terms. Ordinarily, estimated coefficients and the inverse matrix of a new prediction equation will change if either the intercept or a quadratic coefficient is eliminated. This change occurs because covariances between the intercept and pure quadratic terms, and between quadratic terms (nonorthogonality), are not zero. The new estimate of error variance will change because an additional number of degrees of freedom and a reduced number of coefficients are estimated (df $= 20 - 6 = 14$, and $S^2 = 10778.5$).

Variances of predicted values can be estimated by the method of Chapter 4, "3^K Factorial Designs":

$$\text{Var}\,(Y_p) = \mathbf{x}'\,(\mathbf{X}'\mathbf{X})^{-1}\mathbf{x}\,S^2.$$

For example, let the factor levels be $x_1 = 0.5$, $x_2 = -0.5$, and $x_3 = 1.0$, which would be substituted into the prediction equation.

$\mathbf{x}' = (1\quad 0.5\quad 0.25\quad 0.25\quad 1\quad -0.25),$
$Y_p = 5037,$

$$\text{Var}(Y_p) = \mathbf{x}'\begin{bmatrix} 0.167 & 0.000 & -0.056 & -0.056 & -0.056 & 0.000 \\ 0.000 & 0.073 & 0.000 & 0.000 & 0.000 & 0.000 \\ -0.056 & 0.000 & 0.067 & 0.007 & 0.007 & 0.000 \\ -0.056 & 0.000 & 0.007 & 0.067 & 0.007 & 0.000 \\ -0.056 & 0.000 & 0.007 & 0.007 & 0.067 & 0.000 \\ 0.000 & 0.000 & 0.000 & 0.000 & 0.000 & 0.125 \end{bmatrix}\mathbf{x}S^2.$$

$$= (0.108)\,(10778.5) = 1164.1.$$

95% confidence interval: $5037 \pm (2.15)\sqrt{1164.1}$,

$$5037 \pm 73.4.$$

The stationary point is found by taking partial derivatives of the prediction equation with respect to each factor. These derivatives are equated to zero and solved for the factor levels.

$$\frac{\partial Y_p}{\partial X_1} = 143.4 - 143.4x_1 - 71.8x_2 = 0,$$

$$\frac{\partial Y_p}{\partial X_2} = -259.0x_2 - 71.8x_1 = 0,$$

$$\frac{\partial Y_p}{\partial X_3} = -126.2x_3 = 0.$$

The solutions of these equations are given for both coded and uncoded factor levels as

$$x_1 = 1.16, \qquad \text{Enzyme (}\mu\text{g protein)} = 14.6,$$
$$x_2 = -0.32, \qquad \text{amino acid (pmoles)} = 1136,$$
$$x_3 = 0, \qquad \text{pH} = 7.5.$$

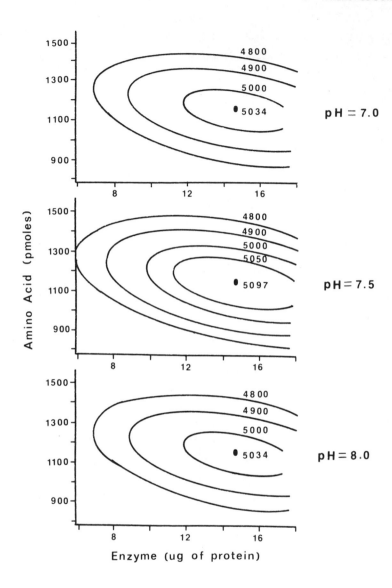

Figure 6.7. Response surface contour plots of arginine-tRNA activity.

The response at the stationary point is $Y_p = 5097$.

The response surface is examined in Figure 6.7 by plotting contour levels for values of X_1 and X_2 at three fixed values of X_3. These contour plots show that a maximum response occurs at the stationary point.

7

Bibliography of Applied Optimization and Response Surface Methods

The articles listed in this bibliography describe specific applications of the optimization and response surface methods discussed in this book. This bibliography is not intended to be all inclusive, but it does contain many examples representative of optimization methods discussed in the text that have appeared in the literature through 1983 and into 1984. Papers are divided into categories representing major fields of interests in chemistry. In each category, papers are listed in inverse chronological order to enable the readers to easily find recent papers in their fields. For references to other methods of optimization and other uses of statistics in chemistry, the reader is urged to consult the April review issues of *Analytical Chemistry*. In 1980 and 1982 these reviews, by B. R. Kowalski, were entitled "Chemometrics." In previous even-numbered years, appearing at two- or four-year intervals, they were under the title of "Statistical and Mathematical Methods in Analytical Chemistry."

7.1 Atomic Absorption/Fluorescence Spectroscopy

Koscielniak, P., and A. Parczewski. 1983. Empirical modeling of the matrix effect in atomic absorption spectrometry: Determination of calcium in presence of aluminum. *Anal. Chim. Acta.* 153:111–119.

Two-factor, two- and three-level factorials with concentrations of calcium and aluminum as the factors.

Michel, R.G., J. Coleman, and J.D. Winefordner. 1978. A reproducible method for preparation and operation of microwave excited electrodeless discharge lamps: SIMPLEX optimization of experimental factors for a cadmium lamp. *Spectrochim. Acta* 33B:195–215.

Procedure for ten variables using detection limit in ppm as response.

Johnson, E.R., C.K. Mann, and T.J. Vickers. 1976. Computer controlled system for study of pulsed hollow cathode lamps. *Appl. Spectrosc.* 30:415–422.

Computer control of lamp variables with closed loop optimization of lamp output by simplex program.

Bentley, G.E., and M.L. Parsons. 1975. Improved method for making electrodeless discharge lamps. *Anal. Lett.* 8:485–501.

Partial optimization by simplex of three factors for making arsine filled lamps with *S/N* ratio as the response.

Parker, L.R., Jr., S.L. Morgan, and S.N. Deming. 1975. Simplex optimization of experimental factors in atomic absorption spectrometry. *Appl. Spectrosc.* 29:429–433.

Simplex study followed by fractional factorial in region of maximum vertex to investigate effects of five variables.

Silvester, M.D., and W.J. McCarthy. 1970. The intensity of electrodeless discharge lamps containing cadmium, mercury, and manganese. *Spectrochim. Acta* 25B:229–243.

Full factorial studies to find optimum operating conditions for three types of lamps, with four factors considered.

Parsons, M.L., and J.D. Winefordner. 1967. Optimization of the critical instrumental parameters for achieving maximum sensitivity and precision in flame-spectrometric methods of analysis. *Appl. Spectrosc.* 21:368–374.

Full factorial and central composite designs for three variables for atomic absorption as well as flame emission spectroscopy.

Cellier, K.M., and H.C.T. Stace. 1966. Determination of optimum operating conditions in atomic absorption spectroscopy. *Appl. Spectrosc.* 20:26–33.

Fractional factorials and central composite designs along with multiple regression analysis for up to seven factors.

7.2   Biological / Clinical Chemistry

Coleman, P.L., J.F. Perry, and J.A. Wehrly. 1983. Optimization of enzyme-based assays in coagulation testing. *Clin. Chem.* 29:603–608.

Kinetic parameters for antithrombin and plasminogen optimized using Box–Behnken design for three factors, with designs repeated for each combination of substrate and buffer.

Fast, D.M., E.J. Sampson, V.S. Whitner, and M. Ali. 1983. Creatine kinase response surfaces explored by use of factorial experiments and simplex maximization. *Clin. Chem.* 29:793–799.

Five-component, five-level central composite response surface study followed by simplex search using centrifugal analyzer.

London, J.W., L.M. Shaw, L. Theodorsen, and J.H. Stromme. 1982. Application of response surface methodology to the assay of *gamma*-glutamyltransferase. *Clin. Chem.* 28:1140–1143.

Central composite design for three factors using centrifugal analyzer.

Burtis, C.A., W.D. Bostick, J.B. Overton, and J.E. Mrochek. 1981. Optimization of a kinetic method by response surface methodology and centrifugal analysis and application to the enzymatic measurement of ethanol. *Anal. Chem.* 53:1154–1159.

Face-centered cube and central composite designs for three factors.

Thompson, J.C., C.T. Hodges, G.L. Dobler, and J.A. Williamson, Jr. 1981. Response-surface-optimized, zinc-enhanced assay for serum alkaline phosphatase. *Clin. Chem.* 27:1171–1175.

Box–Behnken factorial design with four factors at three levels using automatic analyzer.

Rautela, G.S., R.D. Snee, and W.K. Miller. 1979. Response surface co-optimization for reaction conditions in clinical chemical methods. *Clin. Chem* 25:1954–1964.

Various types of factorials including Box–Behnken and face-centered cubes in three enzymatic analyses using automatic analyzers.

Olansky, A.S., and S.N. Deming. 1978. Automated development of a kinetic method for the continuous flow determination of creatinine. *Clin. Chem.* 24:2115–2124.

Simplex optimization for hydroxide and picrate concentrations with response surface mapping by central composite design; automatic analyzer and computer control.

Rautela, G.S., and R.J. Liedtke. 1978. Automated enzymic measurement of total cholesterol in serum. *Clin. Chem.* 24:108–114.

Co-optimization with response surface mapping but specific designs not given.

Olansky, A.S., L.R. Parker, Jr., S.L. Morgan, and S.N. Deming. 1977. Automated development of analytical chemical methods. The determination of serum calcium by the cresolphthalein complexone method. *Anal. Chim. Acta* 95:107–133.

Simplex design for six factors and response surface mapping by central composite design; autoanalyzer and computer control.

Lott, J.A., and K. Turner. 1975. Evaluation of Trinder's glucose oxidase method for measuring glucose in serum and urine. *Clin. Chem.* 21:1754–1760.

Simplex method for four factors using automatic analyzers.

Krause, R.D., and J.A. Lott. 1974. Use of the simplex method to optimize analytical conditions in clinical chemistry. *Clin. Chem.* 20:775–782.

Study of four procedures on autoanalyzer and centrifugal analyzer with up to four factors by enzymatic and spectrophotometric means.

Morgan, S.L., and S.N. Deming. 1974. Simplex optimization of analytical chemical methods. *Anal. Chem.* 46:1170–1181.

Simplex and fractional factorial for five factors in the determination of cholesterol in blood serum.

Rubin, I.B., T.J. Mitchell, and G. Goldstein. 1971. A program of statistical designs for optimizing specific transfer ribonucleic acid assay conditions. *Anal. Chem.* 43:717–721.

Enzymatic method using fractional factorial and central composite designs for up to ten factors.

7.3   Chemiluminescence

Stieg, S., and T.A. Nieman. 1980. Application of a microcomputer-controlled chemiluminescence research instrument to the simultaneous determination of cobalt(II) and silver(I) by gallic acid chemilumines-cence. *Anal. Chem.* 52:800–804.

Modified simplex with three variables was used to find two optima in a multicomponent system.

7.4   Chemistry, General

Harper, S.L., J.F. Walling, D.M. Holland, and L.J. Pranger. 1983. Simplex optimization of multielement ultrasonic extraction of atmo-spheric particulates. *Anal. Chem.* 55:1553–1557.

Quantitative extraction of 13 elements from fiberglass filter was found by simplex procedure with four variables.

Matthews, R.J., S.R. Goode, and S.L. Morgan. 1981. Characterization of an enzymatic determination of arsenic(V) based on response surface methodology. *Anal. Chim. Acta* 133:169–182.

A five-factor simplex study was followed by a Box–Behnken design study to map the response surface.

Otto, M., and G. Werner. 1981. Optimization of a kinetic-catalytic method by use of a numerical model and the simplex method. *Anal. Chim. Acta.* 128:177–183.

Fixed- and variable-size simplexes were used to study rate of decompo-sition of hydrogen peroxide by copper in the presence of pyridine.

McDevitt, R.J., and B.J. Barker. 1980. Simplex optimization of the synergic extraction of a *bis*-diketo copper(II) complex. *Anal. Chim. Acta* 122:223–226.

A study with three factors for the liquid–liquid distribution of copper as a mixed ligand.

Vlacil, F., and H.D. Khanh. 1979. Determination of low concentrations of dibenzyl sulfoxide in aqueous solutions. *Coll. Czech. Chem. Comm.* 44:1908–1917.

Fractional factorials with four variables were followed by a simplex with six variables, with a spectrophotometric endpoint.

Turoff, M.L.H., and S.N. Deming. 1977. Optimization of the extraction of iron(II) from water into cyclohexane with hexafluoroacetylacetone and *tri-n*-butylphosphate. *Talanta.* 24:567–571.

Spectrometric response for a partition experiment by a simplex with four factors followed by a central composite factorial.

Mieling, G.E., R.W. Taylor, L.G. Hargis, J. English, and H.L. Pardue. 1976. Fully automated stopped-flow studies with a hierarchical computer controlled system. *Anal. Chem.* 48:1686–1693.

An automated system using a simplex method to study the titanium–hydrogen peroxide reaction in the presence of EDTA.

Olansky, A.S., and S.N. Deming. 1976. Optimization and interpretation of absorbance response in the determination of formaldehyde with chromotopic acid. *Anal. Chim. Acta* 83:241–249.

Two-variable simplex and factorial studies for a colorimetric method.

Vanroelen, C., R. Smits, P. Van den Winkel, and D.L. Massart. 1976. Application of factor analysis and simplex technique to the optimization of a phosphate determination via molybdenum blue. *Z. Anal. Chem.* 280:21–23.

Three-level factorial and simplex studies with three variables in the extraction and determination of the phosphomolybdate complex.

King, P.G., and S.N. Deming. 1974. UNIPLEX: Single-factor optimization of response in the presence of error. *Anal. Chem.* 46:1476–1481.

Single-factor simplex description for determination of chromate with automatic control of reagent addition.

Czech, F.P. 1973. Simplex optimized acetylacetone method for formaldehyde. *JAOAC* 56:1496–1502.

Three-factor simplexes in a colorimetric procedure.

_____. 1973. Simplex optimized J-acid method for the determination of formaldehyde. *JAOAC* 56:1489–1495.

Colorimetric method using three variables in a simplex.

Houle, M.J., D.E. Long, and D. Smette. 1970. A simplex optimized colorimetric method for formaldehyde. *Anal. Lett.* 3:401–409.

Three-factor simplex for chromotropic acid colorimetric method.

Long, D.E. 1969. Simplex optimization of the response from chemical systems. *Anal. Chim. Acta* 46:193–206.

Details of how to construct a simplex design: two-factor colorimetric method for sulfur dioxide.

Belikov, V.G., N.I. Kokovkin-Shcherbak, and S.Kh. Mutsueva. 1967. Use of statistical methods of planning an experiment to establish the optimum conditions for the determination of antipyrine. *Ind. Lab.* 33:1239–1242. (Translated from *Zavod. Lab.* 33:1049–1052.)

Full factorials and central composite designs for three factors in differential spectrophotometry.

Masalovich, N.S., P.K. Agasyan, V.M. Masalovich, and E.R. Nikolaeva. 1967. Determination of the optimum conditions for oxidation of phosphites with iodine. *Ind. Lab.* 33:1243–1245. (Translated from *Zavod. Lab.* 33:1053–1055.)

Fractional factorials and central composite design for five factors with measurement of unreacted iodine as response.

Box, G.E.P. 1952. Statistical design in the study of analytical methods. *Analyst* 77:879–889.

Factorial designs for four factors in a gravimetric precipitation and filtration procedure.

7.5   Electrochemistry

Cheng, H.-Y., and R.L. McCreery. 1978. Simultaneous determination of reversible potential and rate constant for a first-order EC reaction by potential dependent chronoamperometry. *Anal. Chem.* 50:645–648.

Modified simplex for the two factors, applied potential and rate constant, for benzidine rearrangement reaction.

7.6   Engineering/Industrial Chemistry

Yeh, A.-G., L. Berg, and F.P. McCandless. 1982. A statistical method of designing a catalyst to upgrade solvent refined coal. *ACS Div. Fuel Chem.* Preprints. 27(3–4):63–70.

Full two-level, four-factor (catalysts) factorials to optimize denitrogenation level.

Hendrix, C.D. 1980. Through the response surface with test tube and pipe wrench. *Chemtech.* 10:488–497.

Self-directing optimization by multivariate simplex with product development of a detergent as an example.

_____. 1979. What every technologist should know about experimental design. *Chemtech.* 9:167–174.

Use of factorials and fractional factorials for up to 15 factors in several industrial situations.

7.7   Flame Emission Spectroscopy

Routh, M.W., P.A. Swartz, and M.B. Denton. 1977. Performance of the super modified simplex. *Anal. Chem.* 49:1422–1428.

Comparison of supermodified simplex to modified simplex with two to four factors; computer control with measurement of calcium emission as response.

7.8   Flow Injection Analysis

Betteridge, D., T.J. Sly, A.P. Wade, and J.E.W. Tillman. 1983. Computer-assisted optimization for flow injection analysis of isoprenaline. *Anal. Chem.* 55:1292–1299.

A comparison of the modified simplex procedure for four and five variables with the univariate approach.

Janse, T.A.H.M., P.F.A. Van Der Wiel, and G. Kateman. 1983. Experimental optimization procedures in the determination of phosphate by flow-injection analysis. *Anal. Chim. Acta.* 155:89–102.

Two-level fractional factorials with up to seven factors were followed by full factorials with three factors and then by simplex designs with three factors.

7.9 Gas Chromatography

McMinn, D.G., R.L. Eatherton, and H.H. Hill, Jr. 1984. Multiple-parameter optimization of a hydrogen-atmosphere flame ionization detector. *Anal. Chem.* 56:1293–1298.

Two-level factorial for four and five factors followed by a simplex method for the detection of organo-metallics after gas chromatography.

Leggett, D.J. 1983. Instrumental simplex optimization: Experimental illustrations for an undergraduate laboratory course. *J. Chem. Educ.* 60:707–710.

Two-factor simplex method for separation of alcohols; also a three-factor optimization of an atomic absorption spectrophotometer.

Stuckey, C.L. 1981. A statistical study of a gas chromatographic system equipped with a flame photometric detector. *J. Chromatogr. Sci.* 19:30–34.

Factorials and central composite designs for four factors for analysis of sulfur compounds.

Rubin, I.B., and C.K. Bayne. 1979. Statistical designs for the optimization of the nitrogen-phosphorus gas chromatographic detector response. *Anal. Chem.* 51:541–546.

Factorial and simplex for three variables to maximize nitrogen response and minimize carbon response.

Morgan, S.L., and C.A. Jacques. 1978. Response surface evaluation and optimization in gas chromatography. *J. Chromatogr. Sci.* 16:500–505.

Peak separation for methyl esters with two-level, three-variable factorials for three column loadings.

Yang, F.J., A.C. Brown, III, and S.P. Cram. 1978. Splitless sampling for capillary-column gas chromatography. *J. Chromatogr.* 158:91–109.

Splitless sampling techniques studied using simplex with five factors including residence time of syringe in injector and purge activation delay time.

Holderith, J., T. Toth, and A. Varadi. 1976. Minimizing the time for gas chromatographic analysis. Search for optimal operational parameters by a simplex method. *J. Chromatogr.* 119:215–222.

Simplex for three factors to separate methyl benzenes in minimum length of time.

Morgan, S.L., and S.N. Deming. 1975. Optimization strategies for the development of gas-liquid chromatographic methods. *J. Chromatogr.* 112:267–285.

Simplex for two factors followed by factorial around vertex of maximum response with a chromatographic response function measured.

Van Eenaeme, C., J.M. Bienfait, O. Lambot, and A. Pondant. 1974. Studies on ghosting, an important source of error in the quantitative estimation of free volatile fatty acids by GLC. I. Occurrence of ghosting and factors influencing it. *J. Chromatogr. Sci.* 12:398–403.

_____. 1974. Studies on ghosting, an important source of error in the quantitative estimation of free volatile fatty acids by GLC. II. Effect of sample VFA concentration on ghosting magnitude and effectiveness of remedies to limit ghosting. *J. Chromatogr. Sci.* 12:404–410.

Complex factorial studies for several variables including sequence of sample and ghost eluter injections.

Grant, D.W., and A. Clarke. 1971. A systematic study of the quantitative effects of instrument control on analytical precision in flame ionization gas chromatography. *Anal. Chem.* 43:1951–1957.

The effect of six factors including sample injection procedure using full two-level factorials.

Folmer, O.F., Jr. and D.J. Haase. 1969. A statistical study of gas chromatographic systems employing flame ionization detectors. *Anal. Chim. Acta* 48:63–78.

A comparison of three instruments using fractional factorials for up to five factors.

7.10   Inductively Coupled Plasma Spectroscopy

Montaser, A., G.R. Huse, R.A. Wax, S.-K. Chan, D.W. Golightly,

J.S. Kane, and A.F. Dorrzapf, Jr. 1984. Analytical performance of a low-gas-flow torch optimized for inductively coupled plasma atomic emission spectrometry. *Anal. Chem.* 56:283–288.

Simplex optimization with six factors for simultaneous, multi-element determinations with a weighted signal to background response function.

Leary, J.J., A.E. Brookes, A.F. Dorrzapf, Jr., and D.W. Golightly. 1982. An objective function for optimization techniques in simultaneous multiple-element analysis by inductively coupled plasma spectrometry. *Appl. Spectrosc.* 36:37–40.

Simplex study with two factors using weighted signal to noise response function.

Terblanche, S.P., K. Visser, and P.B. Zeeman. 1981. The modified sequential simplex method of optimization as applied to an inductively coupled plasma source. *Spectrochim. Acta.* 36B:293–297.

A six-factor simplex optimization with detection limits and signal to background ratios as measures of response for copper, lead, and silver determinations.

Ebdon, L., M.R. Cave, and D.J. Mowthorpe. 1980. Simplex optimization of inductively coupled plasmas. *Anal. Chim. Acta.* 115:179–187.

Variable step-size simplex to optimize signal to background ratios for five operating parameters followed by univariate searches.

7.11   Liquid Chromatography

Berridge, J.C. 1984. Automated multiparameter optimization of high-performance liquid chromatographic separations using the sequential simplex procedure. *Analyst.* 109:291–293.

Computer-controlled optimization of three-solvent liquid phase, flow rate, and column temperature for separation of aromatic and heterocyclic acids and bases.

Otto, M., and W. Wegscheider. 1983. Multifactor model for the optimization of selectivity in reversed-phase chromatography. *J. Chromatogr.* 258:11–22.

A three-factor, $6 \times 3 \times 2$ factorial design for six pH values, three methanol concentrations, and two ionic strength values for separating dibasic acids and bases.

Berridge, J.C. 1982. Unattended optimization of reversed-phase high-performance liquid chromatographic separations using the modified simplex algorithm. *J. Chromatogr.* 244:1–14.

Description of computer-controlled procedure for two- and three-component mobile phases and gradient elution profile.

Fast, D.M., P.H. Culbreth, and E.J. Sampson. 1982. Multivariate and univariate optimization studies of liquid-chromatographic separation of steroid mixtures. *Clin. Chem.* 28:444–448.

Variable-size simplexes used to optimize chromatographic response function for a three-variable system including gradient shape.

Wegscheider, W., E.P. Lankmayr, and K.W. Budna. 1982. A chromatographic response function for automated optimization of separations. *Chromatographia* 15:498–504.

Simplex optimization of chromatographic response function with three variables for separation of six polycyclic aromatic hydrocarbons.

Lindberg, W., E. Johansson, and K. Johansson. 1981. Application of statistical optimization methods to the separation of morphine, codeine, noscapine and papaverine in reversed-phase ion-pair chromatography. *J. Chromatogr.* 211:201–212.

Application of full factorial designs and response surface methods for two levels of four factors, with capacity ratio as response.

Sachok, B., J.J. Stranahan, and S.N. Deming. 1981. Two-factor minimum alpha plots for the liquid chromatographic separation of 2,6-disubstituted anilines. *Anal. Chem.* 53:70–74.

A three-level factorial design for separating five components by means of a two-component eluant.

Kong, R.C., B. Sachok, and S.N. Deming. 1980. Combined effects of pH and surface-active-ion concentration in reversed-phase liquid chromatography. *J. Chromatogr.* 199:307–316.

A full two-level four-factor factorial was used to evaluate chromatographic behavior of weak acids and bases and zwitterionic compounds.

Svoboda, V. 1980. Search for optimal eluant composition for isocratic liquid column chromatography. *J. Chromatogr.* 201:241–252.

Computer-controlled modified simplex method for optimal separation time with predetermined maximum column length.

Watson, M.W. and P.W. Carr. 1979. Simplex algorithm for the optimization of gradient elution high-performance liquid chromatography. *Anal. Chem.* 51:1835–1842.

Chromatographic response function measured for five variables for a five-component system of PTH-amino acids.

Rainey, M.L. and W.C. Purdy. 1977. Simplex optimization of the separation of phospholipids by high-pressure liquid chromatography. *Anal. Chim. Acta.* 93:211–219.

Simplex procedure for a five-component sample using a multicomponent eluant.

Smits, R., C. Vanroelen, and D.L. Massart. 1975. The optimization of information obtained by multicomponent chromatographic separation using the simplex technique. *Z. Anal. Chem.* 273:1–5.

Procedure for the separation of several metal ions by a two-factor simplex with informing power as the response.

7.12 Nuclear Magnetic Resonance Spectroscopy

Rubin, I.B. 1984. Determination of optimum pH for the analysis of inorganic phosphate mixtures by 31P nuclear magnetic resonance spectroscopy by a simplex procedure. *Anal. Lett.* 17(A11):1259–1267.

Single-factor simplex optimization for the determination of pyro- and tripolyphosphates in mixtures.

Rubin, I.B., and C.K. Bayne. 1981. Practical application of experimental design methods for optimization of chemical procedures. *Amer. Lab.* 13:51–57.

Two-factor simplex to find minimum solvent signal.

Siegel, M.M. 1981. The use of the modified simplex method for automatic phase correction in Fourier-transform nuclear magnetic resonance spectroscopy. *Anal. Chim. Acta.* 133:103–108.

Two-factor simplex for frequency independent and dependent phase correction angles; computer control.

Cantor, D.M., and J. Jonas. 1976. Automated measurement of spin-lattice relaxation times: Optimized pulsed nuclear magnetic resonance spectrometry. *Anal. Chem.* 48:1904–1906.

Univariate simplex method for pulse length and phase control; computer control.

Ernst, R.R. 1968. Measurement and control of magnetic field homogeneity. *Rev. Sci. Instr.* 39:998–1012.

Steepest descent and simplex methods for two factors for optimization of shimming currents; computer control.

7.13   Neutron Activation Analysis

Burgess, D.D., and P. Hayumbu. 1984. Simplex optimization by advance prediction for single-element instrumental neutron activation analysis. *Anal. Chem.* 56:1440–1443.

Modified simplex techniques used to optimize sample size and irradiation, decay, and counting times to predict gamma-ray spectra of a hypothetical sample.

7.14   Plasma Arc Emission Spectroscopy

Rippetoe, W.E., E.R. Johnson, and T.J. Vickers. 1975. Characterization of the plume of a direct current plasma arc for emission spectrometric analysis. *Anal. Chem.* 47:436–440.

Simplex technique for four variables with signal-to-noise ratio of a calcium line as the response.

7.15   Robotics

Lochmüller, C.H., K.R. Lung and K.R. Cousins. 1985. Applications of optimization strategies in the design of intelligent laboratory robotic procedures. *Anal. Lett.* 18(A4):439–448.

Computer-assisted fixed and variable sized simplex procedures for robotic analysis and optimization of spectrophotometric method with two factors.

7.16   Synthetic Chemistry

Amenta, D.S., C.E. Lamb, and J.J. Leary. 1979. Simplex optimization of yield of *sec*-butylbenzene in a Friedel-Crafts alkylation. *J. Chem. Educ.* 56:557–558.

Two-factor simplex with temperature and catalyst ratio as variables to optimize yield of desired isomer.

Dean, W.K., K.J. Heald, and S.N. Deming. 1975. Simplex optimization of reaction yields. *Science* 189:805–806.

Study of the effects of reaction time and temperature on the yield of a new product.

Kofman, L.P., V.G. Gorskii, B.Z. Brodskii, A.A. Sergo, T.P. Nozdrina, A.I. Osipov, and Yu.V. Nazarov. 1968. Use of optimization methods in developing a process of obtaining prometrine. *Ind. Lab.* 34:87–89. (Translated from *Zavod. Lab.* 34:69–71.)

Use of a five-factor simplex to optimize production of prometrine with determination of sulfur content as response.

7.17   X-Ray Fluorescence Spectroscopy

Jablonski, B.B., W. Wegscheider, and D.E. Leyden. 1979. Evaluation of computer directed optimization for energy dispersive X-ray spectrometry. *Anal. Chem.* 51:2359–2364.

Comparison of simplex and supermodified simplex for current and voltage factors.

Wegscheider, W., B.B. Jablonski, and D.E. Leyden. 1978. Development of an automated procedure for the optimization of multielement analysis with energy dispersive X-ray fluorescence spectroscopy. *Anal. Lett.* A11:27–37.

Two-factor simplex for current and voltage with mapping of iso-response areas; computer control.

REFERENCES

Abramowitz, M., and I.A. Stegun, ed. 1972. *Handbook of mathematical functions with formulas, graphs, and mathematical tables.* New York: Dover.

Beyer, W.H., ed. 1966. *Handbook of tables for probability and statistics.* Cleveland: Chemical Rubber Company.

Bowman, K.O, and L.R. Shenton. 1975. Omnibus test contours for departures from normality based on $\sqrt{b_1}$ and b_2. *Biometrika* 62:243–250.

Box, G.E.P. et al. 1973. Some problems associated with the analysis of multiresponse data. *Technometrics* 15:33–51.

Box, G.E.P., and J.S. Hunter. 1957. Multifactor experimental designs for exploring response surfaces. *Ann. Math. Statist.* 28:195–241.

Box, G.E.P., and G.M. Jenkins. 1970. *Time series analysis: Forecasting and control.* San Francisco: Holden-Day.

Box, G.E.P., and K.B. Wilson. 1951. On the experimental attainment of optimum conditions. *J. Roy. Statist. Soc.* Ser. *B* 13:1–45.

Brooks, S.H., and M.R. Mickey. 1961. Optimum estimation of gradient direction in steepest ascent experiments. *Biometrics* 17:48–56.

Cochran, W.G., and G.M. Cox. 1957. *Experimental designs.* New York: John Wiley.

Cook, R.D., and C.J. Nachtshein. 1980. A comparison of algorithms for constructing exact *D*-optimal designs. *Technometrics* 22:315–324.

Cornell, J.A. 1981. *Experiments with mixtures: Designs, models, and the analysis of mixture data.* New York: John Wiley.

Daniel, C. 1959. Use of half-normal plots in interpreting factorial two-level experiments. *Technometrics* 1:311–341.

Davies, O.L., ed. 1956. *The design and analysis of industrial experiments.* New York: Hafner.

Deming, S.N., and S.L. Morgan. 1983. Teaching the fundamentals of experimental design. *Analytica Chimica Acta* 150:183–198.

Deming, S.N., and L.R. Parker, Jr. 1978. A Review of simplex optimization in analytical chemistry. *CRC Critical Reviews in Analytical Chemistry* 7, no. 3:187–202. Boca Raton, FL: CRC Press.

Dixon, W.J., and F.J. Massey, Jr. 1957. *Introduction to Statistical Analysis.* New York: McGraw-Hill.

Draper, N.R., and H. Smith. 1981. *Applied regression.* New York: John Wiley.

Fisher, R.A. 1925. *The design of experiments.* 4th ed. London: Oliver and Boyd (1947).

Fisher, R.A., and F. Yates. 1953. *Statistical tables for biological, agricultural and medical research.* Edinburgh: Oliver and Boyd.

Graybill, F.A. 1969. *Introduction to matrices with applications in statistics.* Belmont, CA: Wadsworth.

Harrington, Jr., E.C. 1965. The desirability function. *Ind. Quality Control.* 21:494–98.

Hogg, R.V., and A.T. Craig. 1970. *Introduction to mathematical statistics.* New York: Macmillan.

John, P.W.M. 1971. *Statistical design and analysis of experiments.* New York: Macmillan.

Kempthorne, O. 1952. *The design and analysis of experiments.* New York: John Wiley.

Kendall, M.G. and W.R. Buckland. 1982. *A dictionary of statistical terms,* New York: Longman Group.

Khuri, A.I., and M. Conlon. 1981. Simultaneous optimization of multiple responses represented by polynomial regression functions. *Technometrics* 23:363–75.

Kowalski, B.R. 1978. Chemometrics. *Chem. Ind.* London 18:882–84.

Ku, H.H. 1966. Notes on the use of propagation of error formulas. *J. Res. Nat. Bur. Standards* 70C,(4):331/263–341/273.

Lindberg, W., E. Johansson, and K. Johansson. 1981. Application of statistical optimization methods to the separation of morphine, codeine, noscapine and papaverine in reversed-phase ion-pair chromatography. *J. Chromatogr.* 211:201–212.

Long, D.E. 1969. Simplex optimization of the response from chemical systems. *Anal. Chim. Acta* 46:193–206.

Mitchell, T.J. 1974. An algorithm for construction of D-optimal experimental designs. *Technometrics* 16:203–210.

Myers, R.H. 1971. *Response surface methodology.* Boston: Allyn and Bacon.

Nelder, J.A., and R. Mead. 1965. A Simplex method for function minimization. *Computer J.* 7:308–313.

Odeh, R.E. and M. Fox. 1975. *Sample size choice.* New York: Marcel Dekker.

Parzen, E. 1962. *Stochastic processes.* San Francisco: Holden-Day, p. 135.

Pettitt, A.N. 1977. Testing the normality of several independent samples using the Anderson-Darling statistic. *Appl. Statist.* 26(2):156–161.

Plackett, R.L., and J.P. Burman. 1946. The design of optimum multifactorial experiments. *Biometrika* 33:305–325.

Routh, M.W., P.A. Swartz, and M.B. Denton. 1977. Performance of the super modified simplex. *Anal. Chem.* 49:1422–1428.

Ryan, P.B., R.L. Barr, and H.D. Todd. 1980. Simplex techniques for nonlinear optimization. *Anal. Chem.* 52:1460–1467.

Rubin, I.B., T.J. Mitchell, and G. Goldstein. 1971. A program of statistical designs for optimizing specific transfer ribonucleic acid assay conditions. *Anal. Chem.* 43:717–721.

SAS Institute Inc. 1982. *SAS user's guide: Statistics, 1982 edition.* Cary, NC: SAS Institute, Inc.

Sheehan, W.F. 1961. *Physical chemistry.* Boston: Allyn and Bacon, pp. 547–548.

Snedecor, G.W., and W.G. Cochran. 1967. *Statistical methods.* Ames, Iowa: Iowa State University Press.

Spendley, W., G.R. Hext, and F.R. Himsworth. 1962. Sequential application of simplex designs in optimization and evolutionary operation. *Technometrics* 4:441–461.

Steel, R.G.D., and J.H. Torrie. 1960. *Principles and procedures of statistics: with special reference to the biological sciences.* New York: McGraw-Hill.

Weissberg, A., and G.H. Beatty. 1960. Tables of tolerance-limit factors for normal distributions. *Technometrics* 2:483–500.

Welch, W.J. 1982. Branch-and-bound search for experimental designs based on *D*-optimality and other criteria. *Technometrics* 24:41–48.

Zahn, D.A. 1975a. Modifications of and revised critical values for the half-normal plot. *Technometrics* 17:189–200.

_____. 1975b. An empirical study of the half-normal plot. *Technometrics* 17:201–211.

Appendix: BASIC Programs

TABLE A.1. BASIC computer program to calculate the percentile points and t-statistic values for Student's t-distribution.

```
10    REM     ***************************************************************************
20    REM              BASIC PROGRAM TO CALCULATE THE PERCENTILE POINTS
30    REM                  AND T-VALUES FOR THE STUDENT T-DISTRIBUTION.
40    REM     ***************************************************************************
50    '
60    '                          PERCENTILE P0INTS PROGRAM
70    '
80          DATA 3.141593,-0.0953,-0.631,0.81,0.076
90          READ PI,A0,A1,A2,A3
100         PRINT
110         PRINT "DO YOU WANT (1) PERCENTILES OR (2) PROBABILITIES OF T-VALUES?"
120         INPUT "PLEASE ENTER EITHER 1 OR 2 = ",NUM
130         IF NUM = 2 THEN GOTO 440
140         DEF FNA(DF) = 1!/(DF + 1!)
150         DEF FNB(ALPHA) = 1!/SQR(-LOG(ALPHA*(2! - ALPHA)))
160         DEF FNC(DF,ALPHA) = EXP(LOG((SQR(2!*PI)-.5)*ALPHA*SQR(DF))/DF)
170         INPUT "ENTER NUMBER OF DEGREES OF FREEDOM = ",DF
180         INPUT "ENTER SIGNIFICANCE LEVEL = ",ALPHA
190   '
200   '                        APPROXIMATE PERCENTILE POINT
210   '
220         TINV = A0 + A1*FNA(DF) + A2*FNB(ALPHA) + A3*FNC(DF,ALPHA)
230         TVALUE = 1!/TINV
240         IF DF = 1 THEN TVALUE = TAN((1-ALPHA)*PI/2)
```

191

```
250      QUIT = 1
260      GOSUB 510
270      IF ABS(ALPHA + AREA − 1) < .0001 THEN GOTO 380
280   '
290   '                        ADJUST APPROXIMATE PERCENTILE POINT
300   '
310      DELTA1 = .025: DELTA2 = .001
320   WHILE AREA + ALPHA < 1
330      TVALUE = TVALUE + DELTA1: GOSUB 510
340   WEND
350   WHILE AREA + ALPHA > 1
360      TVALUE = TVALUE − DELTA2: GOSUB 510
370   WEND
380      PRINT
390      PRINT "SIGNIFICANCE LEVEL = " ALPHA " DEGREES OF FREEDOM = " DF
400      PRINT
410      LOW = −CINT(1000*TVALUE)/1000: HIGH = CINT(1000*TVALUE)/1000
420      PRINT "PROB[T <= "LOW"] + PROB[T >= "HIGH"] = "ALPHA
430      GOTO 860
440   REM   ***********************************************************************************
450   REM         EVALUATE THE T-DISTRIBUTION AREA FOR A GIVEN T-VALUE
460   REM   ***********************************************************************************
470      INPUT "ENTER NUMBER OF DEGREES OF FREEDOM = ",DF
480      INPUT "ENTER ABSOLUTE VALUE OF T-STATISTIC = ",TVALUE
490      PRINT
500      QUIT = 0
510      THETA = ATN(TVALUE/SQR(DF))
520      TEST = DF MOD 2: IF TEST = 1 THEN GOTO 680
530   '
540   '                        AREA FOR EVEN DEGREES OF FREEDOM
550   '
560      AREA = SIN(THETA)
570      IF DF = 2 GOTO 780
580      SUM = 1!: UP = 0!: DOWN = 0!:
590   FOR K = 1 TO (DF−2)/2
600      DEG = 2*K: UP = LOG(2*K−1)+UP: DOWN = LOG(2*K)+DOWN
610      SUM = SUM + EXP(UP-DOWN)*COS(THETA)^DEG
620   NEXT K
630      AREA = AREA*SUM
640      GOTO 780
650   '
```

```
660  '                   AREA FOR ODD DEGREES OF FREEDOM
670  '
680       AREA = 2*THETA/PI
690       IF DF = 1 THEN GOTO 780
700       SUM = 1!: UP = 0!: DOWN = 0!
710       IF DF = 3 THEN AREA = AREA + 2*SIN(THETA)*COS(THETA)/PI
720       IF DF = 3 THEN GOTO 780
730  FOR K = 1 TO (DF−3)/2
740       DEG = 2*K: UP = LOG(2*K)+UP: DOWN = LOG(2*K + 1)+DOWN
750       SUM = SUM + EXP(UP−DOWN)*COS(THETA)^DEG
760  NEXT K
770       AREA = AREA + 2*SIN(THETA)*COS(THETA)*SUM/PI
780       IF QUIT = 1 THEN RETURN
790       AERA = 1 − AREA
800       PRINT
810       PRINT "NUMBER OF DEGREES OF FREEDOM = " DF
820       PRINT
830       PRINT "PROB[−TVALUE <= T <= TVALUE] = " CINT(1000*AREA)/1000
840       PRINT
850       PRINT "1.0 − PROB[−TVALUE <= T <= TVALUE] = " CINT(1000*AERA)/1000
860  END
```

Table A.2. BASIC computer program to test for normality using Anderson–Darling's statistic and to calculate the skewness and kurtosis statistics.

```
10    REM    *********************************************************************
20    REM              BASIC PROGRAM TO TEST FOR NORMALITY USING
30    REM              THE ANDERSON–DARLING STATISTIC AND THE
40    REM              SKEWNESS AND KURTOSIS STATISTICS.
50    REM    *********************************************************************
60    '
70    '                          MAIN PROGRAM
80    '
90    DIM Y(100), P(100), AD(8)
100   INPUT "ENTER NUMBER OF DATA POINTS = ",N
110   FOR I = 1 TO N
120      INPUT "ENTER DATA VALUE = ",Y(I)
130      SUMY = SUMY + Y(I)
140   NEXT I
150      PRINT "END OF DATA INPUT"
160      AVEY = SUMY/N                       'CALCULATE AVERAGE
170   FOR I = 1 TO N
180      SUMY2 = SUMY2 + (Y(I) − AVEY)^2
190      SUMY3 = SUMY3 + (Y(I) − AVEY)^3
200      SUMY4 = SUMY4 + (Y(I) − AVEY)^4
210   NEXT I
220      STD = SQR(SUMY2/(N−1))             'STANDARD DEVIATION
230      M2 = SUMY2/N                        'SECOND CENTRAL MOMENT
240      M3 = SUMY3/N                        'THIRD CENTRAL MOMENT
250      M4 = SUMY4/N                        'FOURTH CENTRAL MOMENT
260      SKEW = M3/(M2*SQR(M2))             'SKEWNESS
270      KURT = M4/M2^2                      'KURTOSIS
280   REM    *********************************************************************
290   REM                 NORMALIZE AND ORDER THE DATA
300   REM    *********************************************************************
310   FOR H = 1 TO N
320      Y(H) = (Y(H) − AVEY)/STD          'NORMALIZE THE DATA
330   NEXT H
340   FOR J = 1 TO N−1
350      F = 0
360   FOR C = 1 TO N−J
370      IF Y(C+1) >= Y(C) THEN 420
380      T = Y(C)
390      Y(C) = Y(C+1)
400      Y(C+1) = T
```

```
410      F = 1
420   NEXT C
430      IF F = 0 THEN 450
440   NEXT J
450   REM   *********************************************************************
460   REM            CALCULATE NORMAL PERCENTILES OF THE ORDERED DATA
470   REM   *********************************************************************
480      DATA 0.14112821,0.08864027,0.02743349,−0.00039446,0.00328975
490      READ C1,C2,C3,C4,C5
500      DEF FNORM(Z) = 1−1/(1+C1*Z+C2*Z^2+C3*Z^3+C4*Z^4+C5*Z^5)^8
510   FOR J = 1 TO N
520      Z1 = ABS(Y(J))
530      C = 1/SQR(2)
540      P(J) = .5 + .5*FNORM(Z1*C)
550      IF Y(J) < 0 THEN P(J) = 1! − P(J)
560   NEXT J
570   REM   *********************************************************************
580   REM            CALCULATE THE ANDERSON-DARLING STATISTIC
590   REM   *********************************************************************
600      SUMA2 = 0!
610   FOR J = 1 TO N
620      SUMA2 = (2!*J−1)*(LOG(P(J))+LOG(1!−P(N+1−J))) + SUMA2
630   NEXT J
640      ADSTAT = −SUMA2/N − N
650   REM   *********************************************************************
660   REM            CALCULATE PERCENTILE POINTS FOR ANDERSON-DARLING
670   REM   *********************************************************************
680   DATA 0.1674,−0.512,2.10
690   DATA 0.1938,−0.552,1.25
700   DATA 0.2147,−0.608,1.07
710   DATA 0.5597,−0.749,−0.59
720   DATA 0.6305,−0.750,−0.80
730   DATA 0.7514,−0.795,−0.89
740   DATA 0.8728,−0.881,−0.94
750   DATA 1.0348,−1.013,−0.93
760   FOR H = 1 TO 8
770      READ A0,A1,A2
780      AD(H) = A0*(1! + A1/N + A2/N^2)
790   NEXT H
800   REM   *********************************************************************
810   REM                          PRINT RESULTS
820   REM   *********************************************************************
830   PRINT
```

```
840   PRINT "N = " N " MEAN = " AVEY " STANDARD DEVIATION = " STD
850   PRINT
860   PRINT "SKEWNESS = " SKEW " KURTOSIS = " KURT
870   PRINT
880   PRINT "ANDERSON-DARLING STATISTIC = " ADSTAT
890   PRINT
900   PRINT "PERCENTILE POINTS FOR ANDERSON-DARLING STATISTIC"
910   PRINT
920   PRINT "LOWER-TAIL PERCENTILE POINTS"
930   PRINT " 5%:"AD(1) " 10%:"AD(2) " 15%:"AD(3)
940   PRINT
950   PRINT "UPPER-TAIL PERCENTILE POINTS"
960   PRINT " 85%:"AD(4) " 90%:"AD(5) " 95%:"AD(6) " 97.5%:"AD(7) " 99%:"AD(8)
970   END
```

TABLE A.3. BASIC computer program to calculate the plotting positions, guardrails, and final standard deviation for half-normal plots for $M = 15, 31, 63$, and 127 contrasts.

```
10    REM     *******************************************************************

20    REM                    BASIC PROGRAM TO CALCULATE THE PLOTTING

30    REM                    POSITIONS AND GUARDRAILS FOR HALF-NORMAL

40    REM                    PLOTS FOR M = 15, 31, 63, AND 127 CONTRASTS.

50    REM     *******************************************************************

60    '

70    '                              DATA INPUT

80    '

90    DIM Y(130), W(130), Z(130), CRIT(4,2,7), GUARD(4,2,7), R(130)

100     PRINT "ENTER THE SIGNIFICANCE LEVEL FOR DETERMINING THE NONSIGNIFICANT"

110     PRINT "CONTRAST USED FOR THE FINAL ESTIMATE OF THE STANDARD DEVIATION."

120     INPUT "ENTER 1 OR 2 FOR (1) 5% OR (2) 20% = ", IALPHA

130   INPUT "ENTER THE NUMBER OF CONTRASTS = ", M

140   FOR I = 1 TO M

150       INPUT "ENTER CONTRAST VALUE = ", Y(I)

160       W(I) = ABS(Y(I))                'ABSOLUTE VALUE

170   NEXT I

180       PRINT "END OF INPUT DATA"

190   REM     *******************************************************************

200   REM          ORDER ABSOLUTE VALUES AND CALCULATE PERCENTILE POINTS

210   REM     *******************************************************************

220   FOR J = 1 TO M-1

230       F = 0

240   FOR C = 1 TO M-J

250       IF W(C+1) >= W(C) THEN 300

260       T = W(C)

270       W(C) = W(C+1)

280       W(C+1) = T

290       F = 1

300   NEXT C

310       IF F = 0 THEN 330

320   NEXT J

330   REM     *******************************************************************

340   REM                 CALCULATE PERCENTILE POINTS FOR Z-AXIS VALUES

350   REM     *******************************************************************

360       C0 = 2.515517: C1 = .802853: C2 = .010328

370       D1 = 1.432788: D2 = .189269: D3 = .001308

380       DEF FNORM(T) = T-(C0+C1*T+C2*T^2)/(1+D1*T+D2*T^2+D3*T^3)
```

```
390    FOR J = 1 TO M
400        Z1 = .5 − (J−.5)/(2*M)
410        T1 = SQR(LOG(1!/Z1^2))
420        Z(J) = FNORM(T1)
430    NEXT J
440    REM    ***********************************************************************
450    REM                        FIND INITIAL ERROR ESTIMATE
460    REM    ***********************************************************************
470        IF M = 15 THEN WR = W(11): IF M = 15 THEN IDEX = 1
480        IF M = 31 THEN WR = W(22): IF M = 31 THEN IDEX = 2
490        IF M = 63 THEN WR = W(44): IF M = 63 THEN IDEX = 3
500        IF M = 127 THEN WR = W(87): IF M = 127 THEN IDEX = 4
510    REM    ***********************************************************************
520    REM              LOOK UP CRITICAL VALUE AND CALCULATE GUARDRAILS
530    REM    ***********************************************************************
540    FOR JDEX = 1 TO 4
550    FOR ALPHA = 1 TO 2
560    FOR NUM = 1 TO JDEX+3
570        READ CRIT(JDEX,ALPHA,NUM)
580        GUARD(JDEX, ALPHA,NUM) = WR*CRIT(JDEX,ALPHA,NUM)
590    NEXT NUM
600    NEXT ALPHA
610    NEXT JDEX
620    DATA 2.065, 2.427, 2.840, 3.230, 1.533, 1.866, 2.177, 2.470
630    DATA 2.615, 2.807, 2.992, 3.173, 3.351, 2.133, 2.288, 2.439, 2.586, 2.730
640    DATA 3.030, 3.120, 3.209, 3.297, 3.384, 3.470
650    DATA 2.570, 2.647, 2.722, 2.797, 2.872, 2.945
660    DATA 3.396, 3.442, 3.487, 3.532, 3.576, 3.620, 3.663
670    DATA 2.960, 2.999, 3.039, 3.078, 3.116, 3.155, 3.193
680    REM    ***********************************************************************
690    REM              DETERMINE THE NUMBER OF SIGNIFICANT CONTRASTS
700    REM                        AT THE 5% SIGNIFICANCE LEVEL
710    REM    ***********************************************************************
720        SIGNIF = 0: K = 0
730    FOR J = M−IDEX−2 TO M
740        K = K + 1
750        IF W(J) >= GUARD(IDEX,IALPHA,K) THEN SIGNIF = SIGNIF + 1
760    NEXT J
770        NONSIG = M − SIGNIF
780        ENUM = INT(.7*(NONSIG + 1))
```

```
790   REM   ***********************************************************************************
800   REM                    ESTIMATE THE FINAL STANDARD DEVIATION OF
810   REM                              THE ESTIMATE COEFFICIENTS
820   REM   ***********************************************************************************
830        SUM1 = 0!: SUM2 = 0!
840   FOR J = 1 TO ENUM
850        Z1 = .5 - (J-.5)/(2*NONSIG)
860        T1 = SQR(LOG(1!/Z1^2))
870        R(J) = FNORM(T1)
880        SUM1 = SUM1 + W(J)*R(J)
890        SUM2 = SUM2 + R(J)^2
900   NEXT J
910        STD = SUM1/SUM2
920   REM   ***********************************************************************************
930   REM                              PRINT RESULTS
940   REM   ***********************************************************************************
950   PRINT "        HALF — NORMAL PLOT VALUES"
960   PRINT
970   PRINT TAB(9) "ORDER", "|CONTRAST|", "Z-VALUE"
980   FOR J = 1 TO M
990        PRINT TAB(10) J, W(J), CINT(1000*Z(J))/1000
1000  NEXT J
1010  PRINT
1020  PRINT TAB(10) "GUARDRAIL VALUES FOR THE 5% AND 20% SIGNIFICANCE LEVELS"
1030  PRINT
1040  PRINT TAB(10)"NUM", "5% LEVEL", "20% LEVEL"
1050  FOR J = 1 TO IDEX+3
1060        PRINT TAB(10)M-IDEX-3+J, GUARD(IDEX,1,J), GUARD(IDEX,2,J)
1070  NEXT J
1080  PRINT
1090  IF IALPHA = 2 THEN GOTO 1120
1100  PRINT TAB(10) "NUMBER OF NONSIGNIFICANT CONTRASTS AT 5% LEVEL = "; NONSIG
1110  GOTO 1130
1120  PRINT TAB(10) "NUMBER OF NONSIGNIFICANT CONTRASTS AT 20% LEVEL = "; NONSIG
1130  PRINT
1140  PRINT TAB(10) "ESTIMATED STANDARD DEVIATION OF COEFFICIENTS = "; STD
1150  END
```

Index